Recent Advances in Analytical Techniques

(Volume 6)

Edited by

Sibel A. Ozkan

Faculty of Pharmacy
Department of Analytical Chemistry
Ankara University
Ankara, Turkey

Recent Advances in Analytical Techniques

(Volume 6)

Editor: Sibel A. Ozkan

ISSN (Online): 2542-5625

ISSN (Print): 2542-5617

ISBN (Online): 978-981-5124-15-6

ISBN (Print): 978-981-5124-16-3

ISBN (Paperback): 978-981-5124-17-0

First published in 2023.

need for a court order if at any point you breach any terms of this License Agreement. In no event will any delay or failure by Bentham Science Publishers in enforcing your compliance with this License Agreement constitute a waiver of any of its rights.

3. You acknowledge that you have read this License Agreement, and agree to be bound by its terms and conditions. To the extent that any other terms and conditions presented on any website of Bentham Science Publishers conflict with, or are inconsistent with, the terms and conditions set out in this License Agreement, you acknowledge that the terms and conditions set out in this License Agreement shall prevail.

Bentham Science Publishers Pte. Ltd.
80 Robinson Road #02-00
Singapore 068898
Singapore
Email: subscriptions@benthamscience.net

BENTHAM SCIENCE

CONTENTS

PREFACE

This 6th volume of *Recent Advances in Analytical Techniques* contains five comprehensive chapters. The concepts described in this volume reflect the important recent advances in analytical chemistry, including modern quality management aspects of these methods that can find wide use in industry. In addition, the chapters cover important recent trends in analytical methods, including the use of Analytical Techniques for Analysis of Metals and Minerals in Water; Lipidomics Techniques and their Application for Food Nutrition and Health; Recent Advances in the Analysis of Herbicides and Their Transformation Products in Environmental Samples; Nano Porous Anodic Aluminum Oxide: An Overview on Its Fabrication and Potential Applications; PIXE/PIGE Measurements of Archaeological Glass, its Conceptualization and Interpretation: A Case Study. I hope that the readers will greatly enjoy reading the excellent chapters contributed by eminent scientists in their respective fields. I would like to thank all the authors contributing to this volume for their superb contributions.

Also, I would like to thank the Bentham staff, including Ms. Mariam Mehdi (Assistant Manager of Publications), and Mr. Mahmood Alam (Director of Publications) at Bentham Science Publishers for their untiring efforts and efficient interactions with the authors in the publication process.

Sibel A. Ozkan
Faculty of Pharmacy
Department of Analytical Chemistry
Ankara University
Ankara, Turkey

List of Contributors

Harshdeep Kaur — Department of Chemistry, Punjab Agricultural University, Ludhiana, Punjab, India

Jiamin Xu — National R&D Branch Center for Freshwater Aquatic Products Processing Technology (Shanghai), Integrated Scientific Research Base on Comprehensive Utilization Technology for By- Products of Aquatic Product Processing, Ministry of Agriculture and Rural Affairs of the People's Republic of China, Shanghai Engineering Research Center of Aquatic-Product Processing & Preservation, College of Food Science & Technology, Shanghai Ocean University, Shanghai 201306, China

Jiahui Chen — National R&D Branch Center for Freshwater Aquatic Products Processing Technology (Shanghai), Integrated Scientific Research Base on Comprehensive Utilization Technology for By- Products of Aquatic Product Processing, Ministry of Agriculture and Rural Affairs of the People's Republic of China, Shanghai Engineering Research Center of Aquatic-Product Processing & Preservation, College of Food Science & Technology, Shanghai Ocean University, Shanghai 201306, China

Jing Su — Xinhua Hospital, Shanghai Institute for Pediatric Research, Shanghai Key Laboratory of Pediatric Gastroenterology and Nutrition, Shanghai Jiao Tong University, School of Medicine, Shanghai 200092, China

Jian Zhong — National R&D Branch Center for Freshwater Aquatic Products Processing Technology (Shanghai), Integrated Scientific Research Base on Comprehensive Utilization Technology for By- Products of Aquatic Product Processing, Ministry of Agriculture and Rural Affairs of the People's Republic of China, Shanghai Engineering Research Center of Aquatic-Product Processing & Preservation, College of Food Science & Technology, Shanghai Ocean University, Shanghai 201306, China

Makhan Singh Bhullar — Department of Agronomy, Punjab Agricultural University, Ludhiana, Punjab, India

Pervinder Kaur — Department of Agronomy, Punjab Agricultural University, Ludhiana, Punjab, India

Roman Balvanović — Vinča Institute of Nuclear Sciences, National Institute of Serbia, University of Belgrade, Belgrade, Serbia

Saša Đurović — Laboratory of Chromatography, Institute of General and Physical Chemistry, Studetski trg 12, 11158 Belgrade, Serbia

Saša Šorgić — Oenological Laboratory, Heroja Pinkija 49, 26300 Vršac, Serbia

Saša Popov — Oenological Laboratory, Heroja Pinkija 49, 26300 Vršac, Serbia
MS Enviro, Njegoševa 22, 26300 Vršac, Serbia

Snežana Filip — University of Novi Sad, Technical Faculty "Mihajlo Pupin" Zrenjanin, Djure Djakovica b.b., 23000 Zrenjanin, Serbia

Shudan Huang — National R&D Branch Center for Freshwater Aquatic Products Processing Technology (Shanghai), Integrated Scientific Research Base on Comprehensive Utilization Technology for By- Products of Aquatic Product Processing, Ministry of Agriculture and Rural Affairs of the People's Republic of China, Shanghai Engineering Research Center of Aquatic-Product Processing & Preservation, College of Food Science & Technology, Shanghai Ocean University, Shanghai 201306, China

Ting Zhang   National R&D Branch Center for Freshwater Aquatic Products Processing Technology (Shanghai), Integrated Scientific Research Base on Comprehensive Utilization Technology for By- Products of Aquatic Product Processing, Ministry of Agriculture and Rural Affairs of the People's Republic of China, Shanghai Engineering Research Center of Aquatic-Product Processing & Preservation, College of Food Science & Technology, Shanghai Ocean University, Shanghai 201306, China

Ujjal Kumar Sur   Department of Chemistry, Behala College, University of Calcutta, Kolkata, India

Xichang Wang   National R&D Branch Center for Freshwater Aquatic Products Processing Technology (Shanghai), Integrated Scientific Research Base on Comprehensive Utilization Technology for By- Products of Aquatic Product Processing, Ministry of Agriculture and Rural Affairs of the People's Republic of China, Shanghai Engineering Research Center of Aquatic-Product Processing & Preservation, College of Food Science & Technology, Shanghai Ocean University, Shanghai 201306, China

Žiga Šmit   Faculty of Mathematics and Physics, Jožef Stefan Institute, University of Ljubljana, Ljubljana, Slovenia

CHAPTER 1

Analytical Techniques for Analysis of Metals and Minerals in Water

Saša Đurović[1,*], Saša Šorgić[2], Saša Popov[2,3] and Snežana Filip[4]

[1] *Laboratory of Chromatography, Institute of General and Physical Chemistry, Studetski trg 12, 11158 Belgrade, Serbia*

[2] *Oenological Laboratory, Heroja Pinkija 49, 26300 Vršac, Serbia*

[3] *MS Enviro, Njegoševa 22, 26300 Vršac, Serbia*

[4] *University of Novi Sad, Technical Faculty "Mihajlo Pupin" Zrenjanin, Djure Djakovica b.b., 23000 Zrenjanin, Serbia*

Abstract: Investigation of the water samples for content of bulk, trace and heavy metals is of great importance for the humanity. For this purpose, a large number of analytical techniques have been developed. Beside analytical techniques, there are systems and methods for pretreatment and preparation of the samples for analysis. There are also procedures for sampling and sample preservation which are essential for the final result. There are several available instrumental techniques for the analysis of metals in water samples (AAS, GFAAS, ICP-OES, ICP-MS, *etc.*), which can be divided into several groups such as volumetric, spectrophotometric, electrochemical, chromatographic, *etc*. All these techniques may be coupled among themselves and with techniques for sample preparation such as preconcentration techniques. This improves the performance of the applied techniques and decrease the possibility of the contamination of samples. This chapter provides an insight into all these processes and issues from sampling, sample conservation, pretreatment and preparation to the application of different analytical techniques for analysis of water samples.

Keywords: Analysis, Classical methods, Instrumental methods, Metals, Sampling, Sample conservation, Sample pretreatment and preparation, Water.

INTRODUCTION

Water is one of the most precious resources in the world. Contamination of this resource is an important issue to deal with. Presence of the toxic pollutants shows the negative effect on the environment, human health, as well as negative economic effects. Heavy metals in water may originate from natural sources, *i.e.*, eroded sediments, volcanos, *etc.*, or from anthropogenic sources such as waste

---

[*] **Address correspondence to Saša Đurović:** Laboratory of Chromatography, Institute of General and Physical Chemistry, Studetski trg 12, 11158 Belgrade, Serbia; Tel: +381659577200; Email: sasatfns@uns.ac.rs

Sibel A. Ozkan (Ed.)

disposal, industrial effluents, *etc.* These metals may negatively influence the organic life. Their action is connected with their properties, availability, and concentration. Availability depends on the form of these elements, which may be dissolved (dangerous form) and particulated (bounded form in sediments, organic compounds, *etc.*). Balance between these two forms is regulated by pH value and redox potential [1].

It has been reported that the concentration of heavy metals is significantly higher in the populated areas with industry, comparing to their concentration in the wild [2 - 4]. For such reasons, there is a high possibility of contamination of drinking water in these areas followed by an expression of negative effect on human health [5 - 7]. Thus, it is important to develop analytical techniques for monitoring water samples originated from both urban and wild areas. It should be taken into account that these techniques should be able to detect and quantify very low levels of analyzed elements because some elements are present in rather a trace or even ultra-trace levels (μg/L or even ng/L levels). However, it has been mentioned that these levels may be higher in urban areas due to the presence of the industry [8, 9].

Another group of elements is major or bulk elements. Their concentration in the environment is much higher (in mg/L levels). Although they are essential for human health, presence in excessive amounts may lead to different disorders and illnesses [1].

Due to the significance of knowing the levels of all these elements in the water, this chapter's aim is to summarize available methods for sampling, storage, pretreatment, and preparation of water samples for the analysis. Besides, an important task is to present all available analytical techniques for analyzing the metals in both major and trace levels.

SAMPLING, STORAGE, AND PRESERVATION OF THE SAMPLES

Sampling is probably the most important and critical step in all analytical procedures because even a tiny mistake may cause a huge error in obtained result, making the analysis useless. For such reason, sampling procedures need to be followed strictly. Taking the diversity in the nature of the sample itself and concentration of the metals into an account, different sampling methods have been developed [1]. Therefore, the main aspects of the water samples' collection have been defined. It is essential that the collected sample is a representative sample of water which is to be analyzed. To accomplish this, large volumes of water are usually required. Representative samples must be homogenized for preparation of samples for the analysis. It should be also bear in mind that the shorter time between the sampling and analysis is in strong correlation with reliability of the obtained results [10, 11]. It needs to be pointed out that occurrence of the turbidity and/or suspended matter, and application of the methods for their elimination are very important factors in the analytical process [11 - 13]. Chemical profile of the water also has significant influence on the choice of the sampling method. Therefore, a discrete sample should be taken when composition of water is unchanged over the time. However, obtained results showed the composition of the analyzed water at the certain moment. On the other hand, when it comes to the average composition of desired component(s) during the certain period of time, a composite sample should be taken (mix of different samples taken at different period) [14 - 16].

Contamination of the sample during the manipulation is an important factor, which contributes to the final result of the analysis. Magnitude of the concentration of the analyzed element is also a significant contributor. Lower order of magnitude usually means a higher error in the final result. Usual reasons for losing the heavy metals are adsorption on the surface of the storage vessel and/or contamination. Significance of this issue is proven by the available publications on this subject, reporting the necessary steps and precautions to avoid contamination [17 - 19]. Factor, which should be also taken into account, is chemical and biological inertness of the sampling equipment, *i.e.*, used equipment must not change the composition of the water sample. Selected containers must be made of such material that prevents any possible undesirable processes such as adsorption and desorption. Material of the sampling container should be chosen according to the objective of the analysis. If the heavy metals are analyzed, container must not be made of metal in order to prevent contamination of the sample due to the metal leaching. Considering all relevant factors, *e.g.*, sampling efficiency and cost, samples may be kept in a plastic container made of polyethylene or polyvinyl chloride. It is essential that plastic containers are cleaned prior to sample's collection. This could be accomplished by rinsing with diluted hydrochloric acid, distilled and distilled water [20]. Extreme caution is needed when chemical separation is required because chemical reaction may occur, causing the changes in chemical composition of the sample. Such an event happens due to the variations in certain parameters (pH, redox potential, oxygen, *etc.*). In such cases, samples have to be stored at low temperatures (dark place and/or frozen) [1].

When it comes to the bulk elements, such as sodium (Na) and potassium (K), the analyst should be aware that those elements may leach from the glass bottle. Therefore, when bulk elements are to be analyzed, borosilicate or polyethylene vessels should be used. It is also recommended to lower the pH down with nitric acid (pH ≈ 2) in order to prevent the adsorption of these elements on the vessel's wall [21]. Besides, zinc (Zn), manganese (Mn), iron (Fe), and copper (Cu) could also be lost because of the precipitation and/or adsorption. It is necessary to acidify the samples when these elements are to be analyzed. Acidification is accomplished by adding hydrochloric (HCl) or nitric acid (HNO$_3$). Analysis of dissolved fractions requires filtration of the sample through 0.45 μm membrane filters. Dissolved Mn has to be precipitated by oxidation to a higher state with a suitable reagent. However, when total Mn is required, acidification should be used (with HNO$_3$).

For determination of the trace elements, a container made of polypropylene (or high-density polypropylene) and polymers based on fluorinated ethylene are highly recommendable. On the other hand, materials such as soft glass, polyvinyl chloride metals (or plastic-coated metals), rubber, and structural nylon must be avoided [22 - 24]. It is worth mentioning that containers must be carefully cleaned. Cleaning should be performed with a diluted acid solution. Filtration of the samples through the 0.45 μm or 0.20 μm membrane filters is also necessary. Filtration should be performed in an inert atmosphere (N$_2$) for the determination of trace elements [18 - 20]. Ultrafiltration (pore size of 0.001 μm) may also be used for determination of the trace elements [25 - 27]. In the case when filtration *in situ* cannot be performed, it must be done within a few hours after sampling for minimizing the losses of the soluble compounds due to the occurrence of the sorption processes [1].

Immediate analysis of the sample for trace metals is encouraged. However, in the case when it is not possible, the sample must be adequately stored to prevent any possible contamination. Extra care should be dedicated to the possible contamination from the laboratory sources such as distilled water, filters, and containers [28, 29]. In this case, the water sample should be acidified with ultrapure HNO$_3$ (pH < 2). This will prevent the precipitation of the hydroxides or occurrence of the adsorption processes. Sample should also be stored at 4 °C in order to decrease the activity of microorganisms. Refrigeration has several advantages. It does not affect the composition of the samples, does not interfere with applied methods of analysis, and prevents evaporation of certain metals, *i.e.*, mercury, arsenic, selenium, cadmium, and zinc. Addition of the 10% K$_2$Cr$_2$O$_7$ solution is highly recommended for the conservation of the samples when analysis of the mercury is required [1]. As previously mentioned, the addition of the acids into the sample is necessary to prevent both precipitation and sorption processes. Depending on the analyzed elements and chosen analytical technique, preservation can be performed with addition of different acids in different concentrations [30, 31]. After conservation with the acid, samples should be kept in the dark and cold place (4 °C). Analysts should be also aware that time between sampling and analysis must be as short as possible for increasing the reliability of the results [32 - 34]. For obtaining quantitative data about the total recoverable fraction of elements in natural water, including suspended loads, unfiltered samples should be immediately acidified with 1% HNO_3. Aliquot of this mixture is then mixed with concentrated HNO_3 and concentrated HCl, and then heated at 85 °C to reduce the volume. After the heating, sample can be treated with microwaves (closed system) before the analysis [28, 35 - 37]. Contamination issues during the sampling and preservation have been also reported in the case of analysis of the bulk elements. To prevent those issues, acidification and application of the vessels made of borosilicate glass or polyethylene are strongly recommended [38].

PREPARATION OF THE SAMPLES

Preparation of the samples for the analysis includes pretreatment procedures and final preparation for the analysis. Pretreatment is required for the elimination of any possible interferences. The main goal of these procedures is to improve analytical methods and protocols in order to facilitate the analysis of the elements abundant in the trace levels. There are three cases to be considered in the analysis of heavy metals: particulated (suspended), total metals, and dissolved metals. Analytical procedure for determination of total metals requires acidification of the samples (pH $\leq$ 2) before the filtration. In the case of analysis of the dissolved metals, samples should be filtered through the filters (pore size of 0.45 µm), which are previously cleaned with acid in order to prevent contamination of the sample and to remove any particulate matter. For analysis of particulated metals, samples should be filtrated through the previously acidified filters with pore size of 0.45 µm, while the retained matter is to be analyzed [1].

Acidification or stabilization of the samples is a very important and necessary step when it comes to the analysis of the trace elements. Acidification to the pH $\leq$ 1 provides proper conservation of the sample and prevents the occurrence of precipitation and/or sorption. Different acids in different concentrations may be used for this purpose. In most cases, 1% HNO_3 is applied. Another important factor is the storage of the samples. It has been previously mentioned that samples should be kept in a dark and cold place, at 4°C. However, it is strongly recommended to perform analysis as soon as possible after the sampling.

In the case of major elements, filtration is also a necessary and essential step. This is especially important in the case of the samples with high content of particulars and colloids, where filtration forestall sorption and desorption processes. It is worth mentioning that filtration and refrigeration prevents also bacterial activity, which interferes with the analysis. It is of high importance that filters are cleaned with acid solutions prior to application to prevent any possible contamination of the samples [10, 38].

After pretreatment, digestion of the sample should be performed in order to release bonded heavy metals from the organic compounds or complexes. For this purpose, preconcentration and/or separation should be applied. To achieve these goals, coprecipitation [39, 40], complexation and extraction [41, 42], and evaporative methods [43, 44] may be used. Besides previously-mentioned techniques, solid phase extraction is also proven to be effective for the extraction and preconcentration of the metals from water samples [45 - 47], together with the ion-exchange resins [48 - 50]. There are also several reports which have proposed usage of organisms and biomasses for the preconcentration. This proposal is based on the ability of those organisms to absorb the metals [51 - 54]. These steps are followed by the digestion of acids. HNO_3 is the most commonly used acid for this purpose. On the other hand, mixtures of this acid with others (*e.g.*, HCl, H_2SO_4, and $HClO_4$) are also used for digestion. Samples are then evaporated to the lowest possible volume prior to the precipitation, while the addition of HNO_3 is continuing until the clear solution has been obtained. After this step, organic matter in the samples is completely removed.

In particular cases, the dry-ashing method may be used for the analysis of the major elements. For this purpose, samples should be completely evaporated and remain should be converted into an ash in the muffle furnace. Remaining ash is then transferred into a mixture of HNO_3 and hot water. Obtained mixture should be further filtrated and diluted. Moderate digestion is usually used for releasing the ions from their bonded forms. One such example is filtration of the sample through the quartz tube followed by acidification (pH $\leq$ 2) and exposure to UV irradiation (mercury lamp) for digestion [55 - 57]. Hydrogen peroxide (H_2O_2) may also be added to accelerate the degradation of the organic compounds in the water samples [58, 59].

Unlike the trace elements, bulk elements, *i.e.*, K, Na, Ca, and Mg, may be analyzed directly. It has been previously mentioned that trace elements, *e.g.*, transition elements (Mn, Fe, Zn, Cu, *etc.*) require preconcentration prior to analysis. To achieve this, different methods have been developed, *e.g.*, coprecipitation, extraction, and chelation. Coprecipitation is based on addition of metal oxide or organic agent at a defined pH value. Metal ions of interest coprecipitate and, after filtration, may be analyzed directly or be dissolved in acid [40, 60, 61]. In the case of extraction, complexating agent (miscible with water-immiscible organic solvent mixture) is added into a sample of analyzed water. After the extraction, organic layer is separated and analyzed [62, 63]. There are several combination of complexation agents and organic solvent like sodium diethyldithiocarbamate (agent) and methyl isobutyl ketone (solvent) [64, 65].

Third case is chelating by using the solid-phase sorbents, which implies the application of various sorbents for preconcentration of different transition elements. In this case, an important factor is the efficiency of the process, which is determined by sorbent characteristics, *e.g.*, distribution coefficients, stability of the complexes, adsorption and desorption rates, loading ability, selectivity, acid-base behavior, *etc.* [66 - 68]. There are two modes for this method: the column mode and batch mode. Column mode allows automation and direct connection with techniques such as flame atomic absorption spectroscope (FAAS) and inductively coupled plasma (ICP). This approach improves sensitivity and decreases the risk of contamination [69 - 72]. Comparing these methods, it might be concluded that extraction techniques are time-consuming and laborious, while chelating methods decrease preparation time and reduce the possibility of contamination.

ANALYTICAL METHODS

Next step, after the sampling and preparation, is analysis of the prepared samples. Different analytical techniques are available for such purposes. Generally, these techniques may be divided into classical methods, spectrophotometric methods, spectroscopic methods, electrochemical methods, chromatographic methods, and other techniques. In this chapter, all these techniques and methods will be described and summarized.

Conventional Analytical Methods

Conventional analytics mostly has historical importance in these days, but these methods are still applied since they are highly precise and accurate. However, these methods require well-trained chemists to be applied routinely for a large number of samples. The most important methods are volumetric ones. These methods rely on the reaction between known concentration of a titrant (reagent) and the accurately measured volume of the sample. End point of the titration is usually determined by an indicator. Common types of reaction during the titrations are neutralization, oxidation-reductions, complexation, and precipitation. To improve these methods, automatic titrators have been developed. Classic titration techniques are also combined with other methods, such as spectrophotometric, potentiometric, amperometric, which determine the end point and enhance the sensitivity and precision comparing to the visual determination of the end point.

The well-known application of titrimetric techniques is for determination of water hardness (concentration of Ca and Mg), where complexometric methods for Mg analysis have been also developed [73 - 76]. Another alternative method is ion-exchange method. This method showed high reproducibility, but is time-consuming. Combined ion-exchange method with photometric titration for determination of Mg has been reported [77 - 79].

Spectrophotometric Methods

Spectrophotometric methods are based on the creation of the colored compound after reaction with specific reagent. Ultraviolet (UV) or visible (VIS) radiation passes through the sample solution found in the quartz cell. Concentration of the desired analyte is directly proportional to the amount of the absorbed radiation at a certain wavelength [10].

Base on the instrument design, there are single-beam and double-beam spectrophotometers. The single-beam is an older system and offered high sensitivity, while double-beam system offered greater reliability of the obtained results. A single-beam instrument measures the ratio of the incident beam to transmitted beam radiant energy, while the logarithm of a measured ratio represents absorbance of the analyzed system. Single-beam system reads absorbance of the system without sample, and with sample at the light path. This is accomplished by using the beam splitter or pulsed light source. In the case of double-beam instruments, beam splitter splits the incident beam into two beams portions. One of them passes through blank, while the other passes through the analyzed sample. In this case, detector is able to measure the ratio of these two beams in real-time. Both systems are schematically presented in Fig. (1).

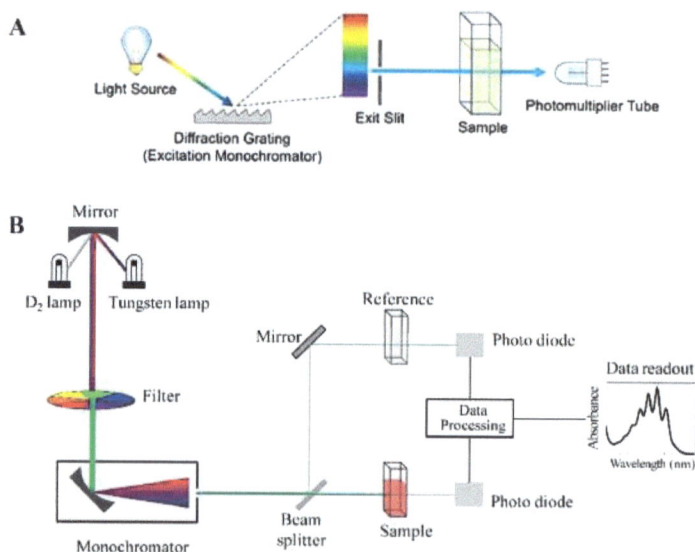

Fig. (1). Principle of the single-beam (**A**) and double-beam (**B**) spectrophotometers.

Spectrophotometric methods are one of the most commonly used methods all over the scientific world and are proven as suitable for the analysis of transition elements in water samples. Flow injection analysis (FIA) has been proposed for the analysis of potassium (K), while combination of sequential injection analysis (SIA) with spectrophotometer as a detector has been proposed for the analysis of Ca [80 - 82]. The FIA method can be used for simultaneous determination of Ca and Mg by using piridylazo resorcinol (PAR) in combination with multivariate calibration [83 - 85].

On the other hand, there are numerous reagents for the determination of transition elements such as Mn, Fe, Zn, and Cu. Besides reagents, there are several different methods for their analysis. One of them is online oxidation-spectrophotometric determination of Mn using FIA [86 - 89]. Analysis of Fe has been also performed routinely regardless of the interference from Cr, Zn, Co, Cu, and Ni in high concentrations [90 - 92]. To facilitate this issue, different techniques have been used, *e.g.*, boiling with acids, the addition of hydroxylamine in excessive amount, liquid-liquid extraction, *etc*. For analysis of Zn different reagents, such as ditizone [93], zincon [94], and xylenol orange [95], are used. There are also several reagents reported for the analysis of Cu in the water samples [96, 97]. Simultaneous method for determination of Cu (II) and Fe (II) using FIA combined with double-beam spectrophotometry has been reported [98, 99]. Besides classical and combined spectrophotometric methods, catalytical spectrophotometric methods (CST) have also been reported. In this method, the transition elements catalyze the reaction among chemical compounds presented in higher concentrations [100 - 104]. The catalytic method combined with FIA for determination of Mn, Al, Cu, Pb and Fe [105 - 109] were developed, where the automatization of the technique enhances the performance, increases the reproducibility, and decreases the time needed for the analysis.

Atomic Absorption Spectroscopy

Atomic absorption spectroscopy (AAS) is one of the most common techniques in the analytical laboratory for determination of metals in different matrixes. It is also one of the cheapest methods, which provides sufficient sensitivity for determination of the major metals in water samples. Certain spectrometers are also able to work in the emission mode. Thus, it is possible to perform an atomic emission spectroscopy analysis, which is more desirable in the case of alkali metals. It is worth mentioning that this technique is free of spectral interferences [1].

If there is a need for increasing the method's sensitivity, simple preconcentration may be performed. Analysts should be aware that the limits of detection (LOD), which have been reported by the manufacturers, have been given for ideal conditions. These LODs are usually compiled to the samples of pure water. In the case of real samples, other ions are presented, which may induce high background. To keep the background in the acceptable ranges, normal connector systems are applied, such as deuterium background correction lamp (D_2BGC) [1].

This technique relies on the absorption of the light of a certain wavelength by the ground-state metals. Metal ions in the water sample are converted to the atomic state after insertion into a flame. When the light of the desired wavelength is supplied by the hollow-cathode lamp or electrodeless discharge lamp, amount of the absorbed light has been measured (Fig. **2**).

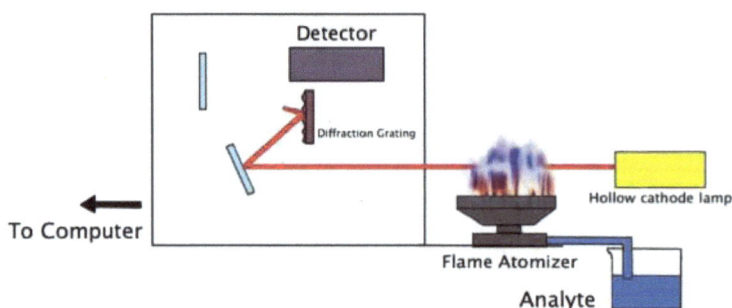

Fig. (2). Schematic representation of the flame atomic absorption spectrometers (FAAS).

Flame atomic absorption spectrometers are of very simple design, while nebulizer is the most complicated and the most essential part of the instrument. The nebulizer (Fig. **3**) converts solution of the analyzed sample into a mist or aerosol, whereupon nebulized sample is introduced into a flame.

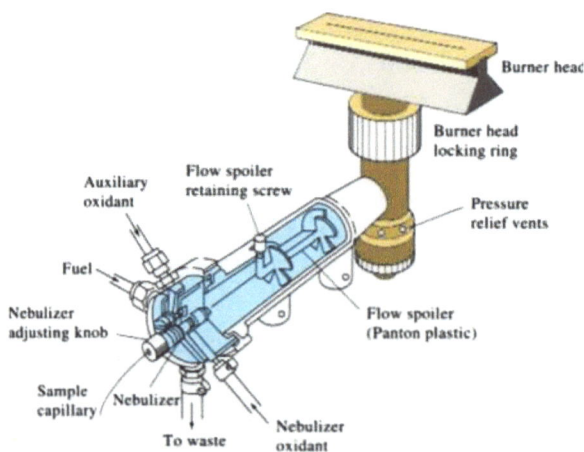

Fig. (3). Schematic representation of the nebulizer for FAAS [110].

After the insertion of the sample into the flame, a specific light source emits the radiation which is focused on the atomic vapor in the flame. This is followed by the entering of the radiation into a monochromator. Output light is then measured by the photomultiplier tube, and an obtained signal is processed by the computer.

Determination of bulk elements (Na, K, Ca, and Mg) usually requires dilution of water samples because they are present in large amounts which exceed the linearity range of the instruments. Combination of FAAS with FIA for automatic determination of these metals has been previously mentioned, together with the application of FIA for insertion of seawater samples into the FAAS in the case of Cu, Zn, Cd, and Pb analysis [1].

Hydride technique has been developed for the analysis of volatile elements, such as As, Sb, Sn, Se, Bi, and Pb. This method enhances the detection limit for these elements 10 to 100-fold. It should be taken into account that these elements are highly toxic and their detection and quantitation are of great importance. Volatile hydrides are generated by adding the acids into a sample and transfer this mixture into a glass vessel with 1% aqueous solution of sodium borohydride ($NaBH_4$). Formed hydride is then introduced by an inert gas into an atomization chamber. Chamber is frequently made of silica and heated in a flame or tube furnace. Hydrides are then decomposed at the high temperature forming atoms. Their concentration is then measured by FAAS [111].

There are other methods proposed to improve the efficiency of the FAAS. One of them is application of the ligands to bond desired metal ions prior to the accumulation step on the adequate sorbent. In this sense, the continuous flow preconcentration technique has been proposed for the analysis of transition metals in water samples [112, 113].

Another AAS technique is electrothermal AAS (ETAAS), also known as graphite furnace AAS (GFAAS) technique. In the GFAAS, 5-100 µL of the sample is transferred into a graphite tube which is open at both ends. This tube fits into a pair of cylindrical graphite electrical contacts at both ends of the tube (Fig. **4**). The contacts are located in a metal housing equipped with water cooling system and with two streams of inert gas. One stream (external) ensures that outside air does not enter the chamber, thus preventing the oxidation of the tube. Second stream (internal) flows into the ends of the tube and out of the central point. This stream has double role. First is to prevent air from entering and the second, most important one, is to carry samples' vapors generated during the heating of the tube. Tube itself is electrically heated in several steps, what is controlled by a program [1, 111].

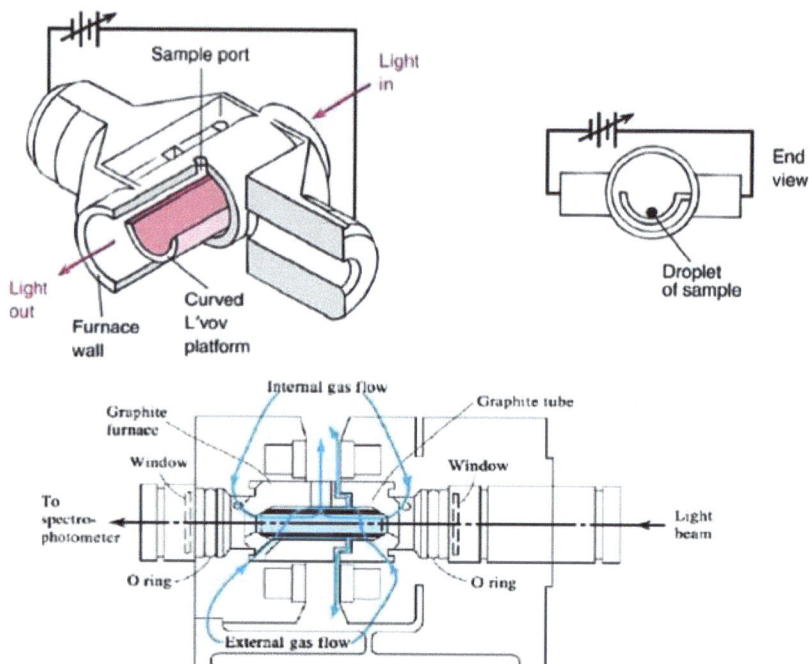

Fig. (4). Cross-sections of the graphite furnace with graphite tube.

During the atomization step graphite furnace heats fast and produces a high density of metal atoms. This ensures much higher selectivity and sensitivity compared to the FAAS technique. However, analytical signal generated in this technique is more sensitive to the composition of the matrix when comparing to the FAAS. Therefore, application of the background corrector system is essential in this technique [114 - 117]. Volatilization of the matrix may be facilitated by using the modifiers (*i.e.*, NH_4NO_3 and $(NH_4)_3PO_4$) [118 - 120]. Addition of the modifiers helps with the elimination of interfering anions such as chlorides. At the same time, they stabilize analytes such as Cd, Pb, and Zn. Insertion of L'vov platform (Fig. **4**) in the graphite furnace ensures the elimination of the interfering ions form the matrix, while reproducibility increases at the same time [121, 122].

There are reports about the application of tungsten (W) instead graphite for manufacturing of the tubes and its application for the determination of several metals [123 - 127]. High selectivity of this technique ensures a direct determination of certain transition metals. (Mn, Cd, Zn, and Fe). However, in certain cases, high background effect occurs which cannot be compensated. In these cases, addition of chelating adsorbents solves this problem by removing the matrix salts and increasing the selectivity through preconcentration of the desired metals [128 - 131]. Furthermore, flow injection accessories for this technique have also been developed [132 - 135]. Ability to better control the possible contamination in the furnace than in the flame under the same experimental conditions is very important advantage of GFAAS over the FAAS [1].

Inductively Coupled Plasma

Inductively coupled plasma (ICP) has proven itself as an excellent technique for the determination of metal ions in water samples. ICP is a partially ionized gas (mostly argon) induced by the quartz torch, where temperature maybe 5,000 to 10,000 K [136 - 138]. This technique shows at least one order of magnitude higher selectivity compared to the FAAS, as well as high stability, excellent reproducibility, and low background level. All these benefits ensure that water samples can be analyzed directly without any preparation and/or pretreatment. The IPC is usually coupled with the optical emission spectrometer (ICP-OES) or mass spectrometer (ICP-MS).

The ICP torch (Fig. **5**) comprises three concentric quartz tubes. Argon gas is passing through those tubes. At the top of this tube is induction coil powered by the RF generator and equipped with the water-cooling system. This generator generates power of 0.5 to 2 kW, while ionization of Ar gas is started by spark from Tesla coil. Induced ions and their electrons interact with the fluctuating magnetic field (Fig. **5**) and give an induction coil, while interaction ensures that ions and electrons flow through the closed annular parts. Existing resistance of the electrons and ions to the flow provides ohmic heating of the plasma [111].

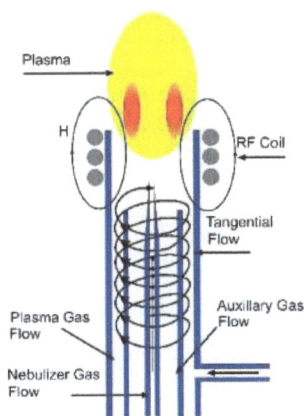

Fig. (5). A typical ICP torch.

The sample is introduced by the peristatic pump to the spry chamber and nebulizer. The Ar caries sample further through the central quartz tube to the torch. Emitted light passes through the entrance window of the Echelle monochromator to the diffraction granting and prism to the CCD detector (Fig. **6**).

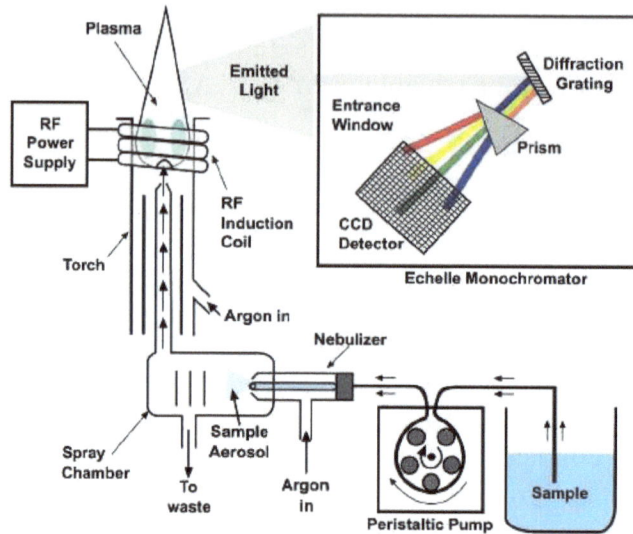

Fig. (6). Schematic presentation of an ICP-OES.

There are two views of the torch in the ICP: radial and axial (Fig. **7**). Thus, it is possible to read both axial and radial in single reading, while vertical torch provides high matrix capability.

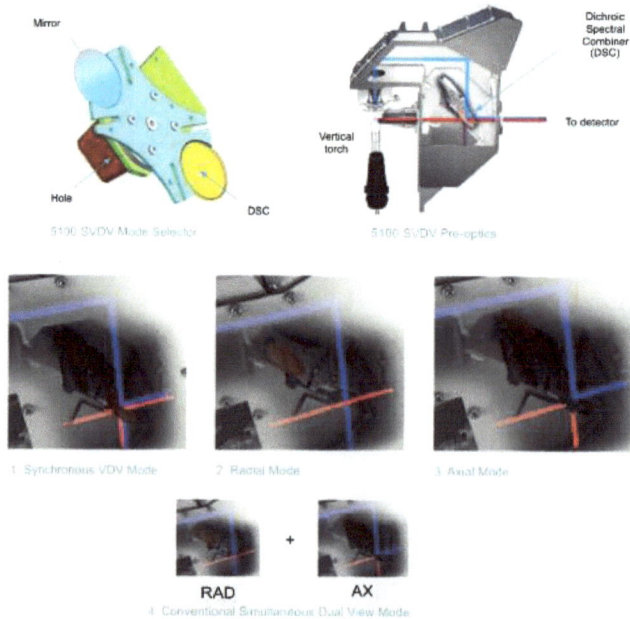

Fig. (7). Cross-section of dichroic spectral combiner and its capabilities in Agilent Technologies 5100 ICP-OES.

The ICP-MS is far more superior to ICP-OES. The common mass analyzer used in this purpose is single quadrupole system (Fig. **8**) [139, 140]. The detection limits for this technique are more than 3-fold lower than those for ICP-OES technique. They are dependent on the sample nature and can be eroded in the case of background increasing or spectral overlapping [141, 142].

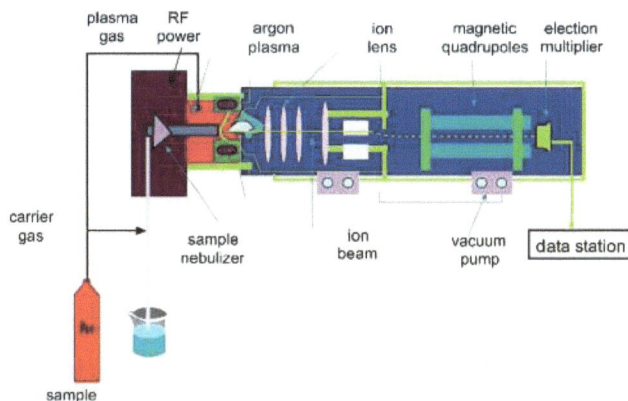

Fig. (8). Cross-section of ICP-MS (single quadrupole) [143].

Both techniques allow simultaneous analysis of a large number of elements. This is highly recommended considering the possibility of investigation of the matrix interferences [144 - 146]. Combination of FIA with ICP and sample enrichment by using the chelating adsorbents can be routinely applied for preconcentration and matrix modification [48, 147 - 153]. Isotope dilution method has been developed and reported for increasing of the analytical performances of the ICP-MS method [154 - 156].

In the end, a combination of the ICP technique with the other analytical techniques, such as ion chromatography (IC-ICP-MS), liquid chromatography (LC-ICP-MS or HPLC-ICP-MS), gas chromatography (GC-ICP-MS), field-flow fractionation (FFF-ICP-MS), capillary electrophoresis (CE-ICP-MS) should be mentioned. Such coupling of the techniques is necessary and essential for increasing the capability for determination of metals in rather trace levels. Some of these methods will soon become standard equipment for determination of toxic elements in analytical laboratories, especially arsenic, cadmium, and mercury. One such coupled instrument is presented in Fig. (**9**).

Electrochemical Methods

Well-known and widely applied techniques belonging to this group are voltammetry and polarography (Fig. **10**). They were used for a long time for analysis of the liquid samples analyzing the dependence of current on applied voltage. Voltage of the polarizable electrode negatively increased in the steps of 1-2 V, while resulting change in current is measured [157, 158]. Voltammetric methods showed high sensitivity, required small volume of the samples, and possibility to work in real-time [159, 160]. It is possible to determine bioavailability of transition metals and complexes which they built with inorganic and organic ligands [161 - 163]. Preconcentration step has also been introduced for improvement of the method's sensitivity. It occurs *in situ* at the surface of working electrode. Such an approach decreases the possibility of sample's contamination [164 - 166].

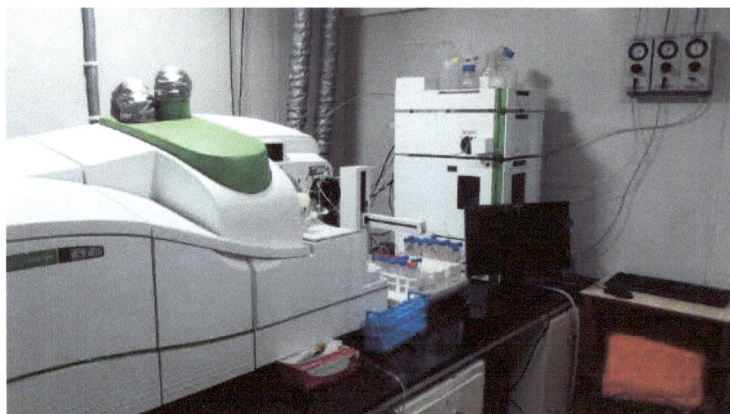

Fig. (9). High performance liquid chromatograph (HPLC) coupled with inductively coupled plasma (ICP) and mass spectrometer (MS).

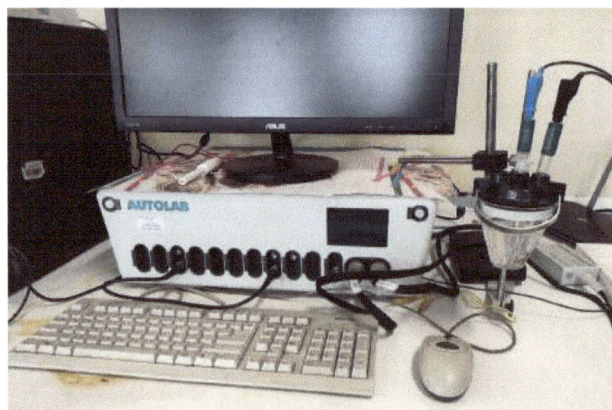

Fig. (10). Metrohm Autolab potentiostat with electrochemical cell for electrochemical analysis.

Another successfully applied method is differential pulse anodic stripping voltammetry (DPASV). In this method, hanging mercury drop (HMD) electrode or rotating glassy carbon thin mercury film (TMF) electrode are used [167 - 170]. It is worth mentioning that TMF electrode ensures significantly higher sensitivity comparing to the HMD electrode. Reason for better performance of the TMF electrode is higher surface-to-volume ratio [171 - 174]. Beside these electrodes, microelectrodes have been developed for this type of analysis [175, 176], while the analysis of metals in seawater has been performed using both cathodic [177 - 180] and anodic [181 - 185] stripping voltammetry. One typical anodic stripping voltammetry curve is given in Fig. (**11**).

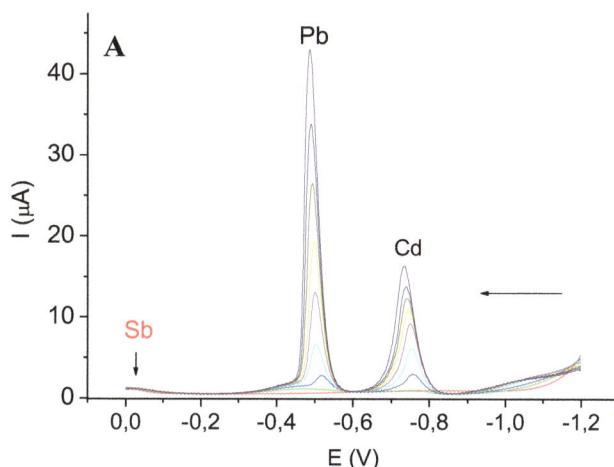

Fig. (11). Typical anodic stripping voltammetry curve for different concentration of lead (Pb) and cadmium (Cd).

Potentiometric methods employ ion-selective electrodes for the analysis of metals in water samples. Solid state electrodes are commonly applied for the determination of sodium and potassium ions or total content of monovalent cations. Different ion selective electrodes have been developed for determination of different metal ions [186 - 190]. Another successfully applied potentiometric technique for water analysis is potentiometric stripping analysis (PSA). This method is insensitive to dissolved oxygen and ensures direct analysis of the samples. This method is very similar to anodic stripping voltammetry, but PSA technique offers the possibility of automation and requires simpler equipment, which is a significant advantage [1].

Chromatographic Techniques

Ion chromatography (IC) is the most frequently used chromatographic technique for the analysis of metals in water. It is one of the most significant methods due to its sensitivity and selectivity, especially because of the possibility to perform analysis of several metals in single run. Besides cations, this technique is able to analyze anions. The typical ion chromatograph is given in Fig. (**12**).

Peak Number	Ret Time	Peak name	Concentration	Area
1	3.65	Lithium	10	6.411e+008
2	4.33	Sodium	10	2.076e+008
3	4.95	Ammonium	10	1.047e+008
4	6.27	Potassium	10	1.186e+008
5	9.25	Magnesium	10	3.161e+008
6	11.50	Calcium	10	1.732e+008
7	13.43	Strontium	10	7.174e+007
8	20.45	Barium	10	4.464e+007

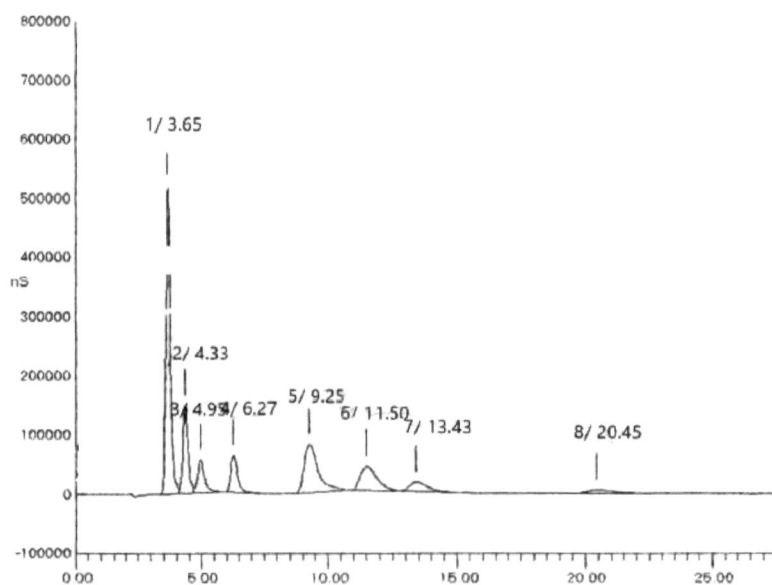

Fig. (12). Typical ion chromatogram of mixture of several cations with their designation.

This technique relies on the separation of metal ions in the column with low-capacity cation-exchange resin as a stationary phase. System may be single-column or suppressed one. Both ones are used for analysis of major cations in water samples [191 - 193]. Different detectors have been used for this system. However, conductometric detector has been proven as the best solution [194 - 197].

Besides ion chromatography, other chromatographic techniques may be used. Hence, thin layer chromatography (TLC) has been used as metal-ion separation technique at first [198 - 200]. It has been also applied for prediction of metal complexes behavior in HPLC systems [201 - 203], which is another chromatographic technique suitable for analysis of metal ions and their complexes [204 - 206]. As previously mentioned, the HPLC techniques may be coupled with ICP-MS or AAS to improve the performance of the analytical techniques.

Other Techniques

Metal ions may react with the suitable ligands resulting in the formation of the complex compounds, which are fluorescent or phosphorescent. Photoluminescence may be applied in direct or indirect mode. Direct mode includes the creation of the fluorescence compound with the analyte. On the other hand, indirect mode is based on the measuring of analyte's influence on the luminescence of other species [1]. There are also several reagents which may increase analytical performances (sensitivity and selectivity) of the method [207, 208]. Pulse laser has been reported to increase signal-to-noise (S/N) ratio [1, 209]. Instrumentation for this technique is quite simple, consisting of reactor and photomultiplier, which is a significant advantage.

X-ray fluorescence spectrometry (XRFS) is also one of the well-known methods for analysis of the metals in aqueous samples. Preconcentration of the samples is performed by using a chelating ion exchanger (column packing or membrane form) [210, 211]. Besides chelating ion exchanger, solvent extraction may be also used for preconcentration of the samples prior to the analysis [212, 213].

CONCLUSION

Water is the most significant resource these days. Apart from the human necessity for water, there is continuous increasing demands by the industries, especially food and pharmaceutical. Importance of high quality water and strict regulations forced analytical industry to develop different analytical techniques which are able to fulfill these requirements. As a consequence, market offers different techniques for the analysis of the water samples, such as atomic absorption spectrometry, inductively coupled plasma, chromatographic techniques, *etc*. New trend is coupling of these techniques among themselves with addition of more powerful detectors such as mass spectrometer. It may be concluded that analytical equipment has recorded significant development over the past 70 years from classic volumetric methods to powerful instrumental techniques, which offer very low detection limits, high sensitivity and selectivity.

REFERENCES

[1]   Marcovecchio, J.E.; Botte, S.E.; Freije, R.H. Heavy Metals, Major Metals, Trace Elements. In: *Handbook of Water Analysis 2nd* Nollet, L., Ed.; CRC Press: Boca Raton, Florida, USA, **2007**; pp. 275-311.

[2]   Breward, N. Heavy-metal contaminated soils associated with drained fenland in Lancashire, England, UK, revealed by BGS Soil Geochemical Survey. *Appl. Geochem.,* **2003**, *18*, 1663-1670.
[http://dx.doi.org/10.1016/S0883-2927(03)00081-7]

[3]   Marcovecchio, J.E. The use of Micropogonias furnieri and Mugil liza as bioindicators of heavy metals pollution in La Plata river estuary, Argentina. *Sci. Total Environ.,* **2004**, *323*(1-3), 219-226.
[http://dx.doi.org/10.1016/j.scitotenv.2003.09.029] [PMID: 15081729]

[4] Rawat, M.; Moturi, M.C.Z.; Subramanian, V. Inventory compilation and distribution of heavy metals in wastewater from small-scale industrial areas of Delhi, India. *J. Environ. Monit.,* **2003**, *5*(6), 906-912.
[http://dx.doi.org/10.1039/b306628b] [PMID: 14710931]

[5] Quinn, M.R.; Feng, X.; Folt, C.L.; Chamberlain, C.P. Analyzing trophic transfer of metals in stream food webs using nitrogen isotopes. *Sci. Total Environ.,* **2003**, *317*(1-3), 73-89.
[http://dx.doi.org/10.1016/S0048-9697(02)00615-0] [PMID: 14630413]

[6] Obbard, J.P. Ecotoxicological assessment of heavy metals in sewage sludge amended soils. *Appl. Geochem.,* **2001**, *16*, 1405-1411.
[http://dx.doi.org/10.1016/S0883-2927(01)00042-7]

[7] Jha, A.N. Genotoxicological studies in aquatic organisms: an overview. *Mutat. Res.,* **2004**, *552*(1-2), 1-17.
[http://dx.doi.org/10.1016/j.mrfmmm.2004.06.034] [PMID: 15352315]

[8] Vidal, J.; Pérez-Sirvent, C.; Martínez-Sánchez, M.J.; Navarro, M.C. Origin and behaviour of heavy metals in agricultural Calcaric Fluvisols in semiarid conditions. *Geoderma,* **2004**, *121*, 257-270.
[http://dx.doi.org/10.1016/j.geoderma.2003.12.001]

[9] Nriagu, J.O.; Pacyna, J.M. Quantitative assessment of worldwide contamination of air, water and soils by trace metals. *Nature,* **1988**, *333*(6169), 134-139.
[http://dx.doi.org/10.1038/333134a0] [PMID: 3285219]

[10] Kremling, K. Determination of trace elements, In: Methods of Seawater Analysis. , **1999**; pp. 270-326.
[http://dx.doi.org/10.1002/9783527613984.ch12]

[11] Hashemi, P. Major Elements. In: *Handbook of Water Analysis*; Nollet, L.M.L., Ed.; Marcel Dekker: New York, **2000**; pp. 409-438.

[12] Eom, I.Y.; Dasgupta, P.K. Frequency-selective absorbance detection: Refractive index and turbidity compensation with dual-wavelength measurement. *Talanta,* **2006**, *69*(4), 906-913.
[http://dx.doi.org/10.1016/j.talanta.2005.11.033] [PMID: 18970656]

[13] Gagnon, C.; Saulnier, I. Distribution and fate of metals in the dispersion plume of a major municipal effluent. *Environ. Pollut.,* **2003**, *124*(1), 47-55.
[http://dx.doi.org/10.1016/S0269-7491(02)00433-5] [PMID: 12683982]

[14] Gonzalez-Soto, E.; Alonso-Rodrıguez, E.; Prada-Rodrıguez, D. Heavy metals. In: *Handbook of Water Analysis*; Nollet, L.M.L., Ed.; Marcel Dekker: New York, **2000**; pp. 439-457.

[15] Kohler, E.A.; Poole, V.L.; Reicher, Z.J.; Turco, R.F. Nutrient, metal, and pesticide removal during storm and nonstorm events by a constructed wetland on an urban golf course. *Ecol. Eng.,* **2004**, *23*, 285-298.
[http://dx.doi.org/10.1016/j.ecoleng.2004.11.002]

[16] Harvey, J.W.; Newlin, J.T.; Krupa, S.L. Modeling decadal timescale interactions between surface water and ground water in the central Everglades, Florida, USA. *J. Hydrol. (Amst.),* **2006**, *320*, 400-420.
[http://dx.doi.org/10.1016/j.jhydrol.2005.07.024]

[17] Petersen, W.; Geisler, C-D.; Schroeder, F.; Knauth, H-D. AISIT — a new device for remote-controlled sampling of dissolved and particle-bound trace elements in surface waters. *J. Sea Res.,* **1998**, *40*, 179-191.
[http://dx.doi.org/10.1016/S1385-1101(98)00023-9]

[18] Capodaglio, G.; Barbante, C.; Turetta, C.; Scarponi, G.; Cescon, P. Analytical quality control: Sampling procedures to detect trace metals in environmental matrices. *Mikrochim. Acta,* **1996**, *123*, 129-136.
[http://dx.doi.org/10.1007/BF01244386]

[19] Moliner-Martínez, Y.; Meseguer-Lloret, S.; Tortajada-Genaro, L.A.; Campíns-Falcó, P. Influence of water sample storage protocols in chemiluminescence detection of trace elements. *Talanta,* **2003**, *60*(2-3), 257-268.
[http://dx.doi.org/10.1016/S0039-9140(03)00068-7] [PMID: 18969048]

[20] Yunes, N.; Moyano, S.; Cerutti, S.; Gásquez, J.A.; Martinez, L.D. On-line preconcentration and determination of nickel in natural water samples by flow injection-inductively coupled plasma optical emission spectrometry (FI-ICP-OES). *Talanta,* **2003**, *59*(5), 943-949.
[http://dx.doi.org/10.1016/S0039-9140(02)00639-2] [PMID: 18968983]

[21] Alfassi, Z.B.; Wai, C.M. *Preconcentration Techniques for Trace Elements*; CRC press, **1992**.

[22] Hall, G.E.M.; Bonham-Carter, G.F.; Horowitz, A.J.; Lum, K.; Lemieux, C.; Quemerais, B. The effect of using different 0.45 μm filter membranes on 'dissolved' element concentrations in natural waters. *Appl. Geochem.,* **1996**, *11*, 243-249.
[http://dx.doi.org/10.1016/0883-2927(96)00059-5]

[23] Viers, J.; Dupré, B.; Polvé, M.; Schott, J.; Dandurand, J-L.; Braun, J-J. Chemical weathering in the drainage basin of a tropical watershed (Nsimi-Zoetele site, Cameroon): comparison between organic-poor and organic-rich waters. *Chem. Geol.,* **1997**, *140*, 181-206.
[http://dx.doi.org/10.1016/S0009-2541(97)00048-X]

[24] Morford, J.L.; Emerson, S.R.; Breckel, E.J.; Kim, S.H. Diagenesis of oxyanions (V, U, Re, and Mo) in pore waters and sediments from a continental margin. *Geochim. Cosmochim. Acta,* **2005**, *69*, 5021-5032.
[http://dx.doi.org/10.1016/j.gca.2005.05.015]

[25] Konhauser, K.O.; Powell, M.A.; Fyfe, W.S.; Longstaffe, F.J.; Tripathy, S. Trace element chemistry of major rivers in Orissa State, India. *Environ. Geol.,* **1997**, *29*, 132-141.
[http://dx.doi.org/10.1007/s002540050111]

[26] Kimball, B.A.; Callender, E.; Axtmann, E.V. Effects of colloids on metal transport in a river receiving acid mine drainage, upper Arkansas River, Colorado, U.S.A. *Appl. Geochem.,* **1995**, *10*, 285-306.
[http://dx.doi.org/10.1016/0883-2927(95)00011-8]

[27] Barringer, J.L.; Szabo, Z.; Schneider, D.; Atkinson, W.D.; Gallagher, R.A. Mercury in ground water, septage, leach-field effluent, and soils in residential areas, New Jersey coastal plain. *Sci. Total Environ.,* **2006**, *361*(1-3), 144-162.
[http://dx.doi.org/10.1016/j.scitotenv.2005.05.037] [PMID: 15996719]

[28] Uchida, H.; Ito, T. Analytical Characteristics of Inductively Coupled Oxygen-Plasma Mass Spectrometry Assisted by Adding Argon to the Outer Gas. *Anal. Sci.,* **1997**, *13*, 391-396.
[http://dx.doi.org/10.2116/analsci.13.391]

[29] Zhang, J-Z.; Fischer, C.J.; Ortner, P.B. Laboratory glassware as a contaminant in silicate analysis of natural water samples. *Water Res.,* **1999**, *33*, 2879-2883.
[http://dx.doi.org/10.1016/S0043-1354(98)00508-9]

[30] Yu, L-P.; Yan, X-P. Factors affecting the stability of inorganic and methylmercury during sample storage. *TrAC. Trends Analyt. Chem.,* **2003**, *22*, 245-253.
[http://dx.doi.org/10.1016/S0165-9936(03)00407-2]

[31] Batley, G.E. *Trace Element Speciation: Analytical Methods and Problems*; CRC press, **1990**.

[32] Broekaert, J.A.C.; Siemens, V. Recent trends in atomic spectrometry with microwave-induced plasmas. *Spectrochim. Acta B At. Spectrosc.,* **2004**, *59*, 1823-1839.
[http://dx.doi.org/10.1016/j.sab.2004.08.006]

[33] Paukert, T.; Sirotek, Z. A study of the microwave treatment of water samples from the Elbe River, Bohemia, Czech Republic. *Chem. Geol.,* **1993**, *107*, 133-144.
[http://dx.doi.org/10.1016/0009-2541(93)90106-S]

[34]   Hsu, S-Y.; Kao, H-Y. Effects of storage conditions on chemical and physical properties of electrolyzed oxidizing water. *J. Food Eng.*, **2004**, *65*, 465-471.
[http://dx.doi.org/10.1016/j.jfoodeng.2004.02.009]

[35]   Hashemi, P.; Olin, A. Equilibrium and kinetic properties of a fast iminodiacetate based chelating ion exchanger and its incorporation in a FIA-ICP-AES system. *Talanta*, **1997**, *44*(6), 1037-1053.
[http://dx.doi.org/10.1016/S0039-9140(96)02189-3] [PMID: 18966835]

[36]   Harzdorf, C.; Janser, G.; Rinne, D.; Rogge, M. Application of microwave digestion to trace organoelement determination in water samples. *Anal. Chim. Acta*, **1998**, *374*, 209-214.
[http://dx.doi.org/10.1016/S0003-2670(98)00511-X]

[37]   Acar, O. Determination of cadmium, copper and lead in soils, sediments and sea water samples by ETAAS using a Sc + Pd + NH$_{(4)}$NO$_{(3)}$ chemical modifier. *Talanta*, **2005**, *65*(3), 672-677.
[http://dx.doi.org/10.1016/j.talanta.2004.07.035] [PMID: 18969851]

[38]   McCleskey, R.B.; Nordstrom, D.K.; Maest, A.S. Preservation of water samples for arsenic(III/V) determinations: an evaluation of the literature and new analytical results. *Appl. Geochem.*, **2004**, *19*, 995-1009.
[http://dx.doi.org/10.1016/j.apgeochem.2004.01.003]

[39]   Saracoglu, S.; Soylak, M.; Elci, L. Separation/preconcentration of trace heavy metals in urine, sediment and dialysis concentrates by coprecipitation with samarium hydroxide for atomic absorption spectrometry. *Talanta*, **2003**, *59*(2), 287-293.
[http://dx.doi.org/10.1016/S0039-9140(02)00501-5] [PMID: 18968910]

[40]   Soylak, M.; Saracoglu, S.; Divrikli, U.; Elci, L. Coprecipitation of heavy metals with erbium hydroxide for their flame atomic absorption spectrometric determinations in environmental samples. *Talanta*, **2005**, *66*(5), 1098-1102.
[http://dx.doi.org/10.1016/j.talanta.2005.01.030] [PMID: 18970095]

[41]   Howard, J.L.; Vandenbrink, W.J. Sequential extraction analysis of heavy metals in sediments of variable composition using nitrilotriacetic acid to counteract resorption. *Environ. Pollut.*, **1999**, *106*(3), 285-292.
[http://dx.doi.org/10.1016/S0269-7491(99)00115-3] [PMID: 15093024]

[42]   Agbenin, J.O.; De Abreu, C.A.; van Raij, B. Extraction of phytoavailable trace metals from tropical soils by mixed ion exchange resin modified with inorganic and organic ligands. *Sci. Total Environ.*, **1999**, *227*, 187-196.
[http://dx.doi.org/10.1016/S0048-9697(99)00027-3]

[43]   de Lima, T.S.; Campos, P.C.; Afonso, J.C. Metals recovery from spent hydrotreatment catalysts in a fluoride-bearing medium. *Hydrometallurgy*, **2005**, *80*, 211-219.
[http://dx.doi.org/10.1016/j.hydromet.2005.08.009]

[44]   Yanagimachi, I.; Nashida, N.; Iwasa, K.; Suzuki, H. Enhancement of the sensitivity of electrochemical stripping analysis by evaporative concentration using a super-hydrophobic surface. *Sci. Technol. Adv. Mater.*, **2005**, *6*, 671-677.
[http://dx.doi.org/10.1016/j.stam.2005.06.017]

[45]   Liu, Y.; Chang, X.; Wang, S.; Guo, Y.; Din, B.; Meng, S. Solid-phase extraction and preconcentration of cadmium(II) in aqueous solution with Cd(II)-imprinted resin (poly-Cd(II)-DAAB-VP) packed columns. *Anal. Chim. Acta*, **2004**, *519*, 173-179.
[http://dx.doi.org/10.1016/j.aca.2004.06.017]

[46]   Hejazi, L.; Mohammadi, D.E.; Yamini, Y.; Brereton, R.G. Solid-phase extraction and simultaneous spectrophotometric determination of trace amounts of Co, Ni and Cu using partial least squares regression. *Talanta*, **2004**, *62*(1), 183-189.
[http://dx.doi.org/10.1016/S0039-9140(03)00412-0] [PMID: 18969279]

[47]   Martínez-Barrachina, S.; del Valle, M. Use of a solid-phase extraction disk module in a FI system for

the automated preconcentration and determination of surfactants using potentiometric detection. *Microchem. J.,* **2006**, *83*, 48-54.
[http://dx.doi.org/10.1016/j.microc.2006.01.022]

[48] Jiménez, M.S.; Velarte, R.; Castillo, J.R. Performance of different preconcentration columns used in sequential injection analysis and inductively coupled plasma-mass spectrometry for multielemental determination in seawater. *Spectrochim. Acta B At. Spectrosc.,* **2002**, *57*, 391-402.
[http://dx.doi.org/10.1016/S0584-8547(01)00401-3]

[49] Hirata, S.; Kajiya, T.; Takano, N.; Aihara, M.; Honda, K.; Shikino, O. Determination of trace metals in seawater by on-line column preconcentration inductively coupled plasma mass spectrometry using metal alkoxide glass immobilized 8-quinolinol. *Anal. Chim. Acta,* **2003**, *499*, 157-165.
[http://dx.doi.org/10.1016/S0003-2670(03)00949-8]

[50] Çekiç, S.D.; Filik, H.; Apak, R. Use of an o -aminobenzoic acid-functionalized XAD-4 copolymer resin for the separation and preconcentration of heavy metal(II) ions. *Anal. Chim. Acta,* **2004**, *505*, 15-24.
[http://dx.doi.org/10.1016/S0003-2670(03)00211-3]

[51] Hawari, A.H.; Mulligan, C.N. Biosorption of lead(II), cadmium(II), copper(II) and nickel(II) by anaerobic granular biomass. *Bioresour. Technol.,* **2006**, *97*(4), 692-700.
[http://dx.doi.org/10.1016/j.biortech.2005.03.033] [PMID: 15935654]

[52] Naja, G.; Volesky, B. Multi-metal biosorption in a fixed-bed flow-through column. *Colloids Surf. A Physicochem. Eng. Asp.,* **2006**, *281*, 194-201.
[http://dx.doi.org/10.1016/j.colsurfa.2006.02.040]

[53] Iyer, A.; Mody, K.; Jha, B. Biosorption of heavy metals by a marine bacterium. *Mar. Pollut. Bull.,* **2005**, *50*(3), 340-343.
[http://dx.doi.org/10.1016/j.marpolbul.2004.11.012] [PMID: 15757698]

[54] Zouboulis, A.; Loukidou, M.; Matis, K. Biosorption of toxic metals from aqueous solutions by bacteria strains isolated from metal-polluted soils. *Process Biochem.,* **2004**, *39*, 909-916.
[http://dx.doi.org/10.1016/S0032-9592(03)00200-0]

[55] Achterberg, E.P.; van den Berg, C.M.G. In-line ultraviolet-digestion of natural water samples for trace metal determination using an automated voltammetric system. *Anal. Chim. Acta,* **1994**, *291*, 213-232.
[http://dx.doi.org/10.1016/0003-2670(94)80017-0]

[56] Bowie, A.R.; Whitworth, D.J.; Achterberg, E.P.; Mantoura, R.F.C.; Worsfold, P.J. Biogeochemistry of Fe and other trace elements (Al, Co, Ni) in the upper Atlantic Ocean. *Deep Sea Res. Part I Oceanogr. Res. Pap.,* **2002**, *49*, 605-636.
[http://dx.doi.org/10.1016/S0967-0637(01)00061-9]

[57] Achterberg, E.P.; van den Berg, C.M.G.; Colombo, C. High resolution monitoring of dissolved Cu and Co in coastal surface waters of the Western North Sea. *Cont. Shelf Res.,* **2003**, *23*, 611-623.
[http://dx.doi.org/10.1016/S0278-4343(03)00003-7]

[58] Sáenz, V.; Blasco, J.; Gómez-Parra, A. Speciation of heavy metals in recent sediments of three coastal ecosystems in the Gulf of Cádiz, southwest Iberian Peninsula. *Environ. Toxicol. Chem.,* **2003**, *22*(12), 2833-2839.
[http://dx.doi.org/10.1897/02-448] [PMID: 14713021]

[59] Mucha, A.P.; Vasconcelos, M.T.S.D.; Bordalo, A.A. Spatial and seasonal variations of the macrobenthic community and metal contamination in the Douro estuary (Portugal). *Mar. Environ. Res.,* **2005**, *60*(5), 531-550.
[http://dx.doi.org/10.1016/j.marenvres.2004.12.004] [PMID: 15919109]

[60] Eltayeb, M.A.H.; Van Grieken, R.E. Coprecipitation with aluminium hydroxide and x-ray fluorescence determination of trace metals in water. *Anal. Chim. Acta,* **1992**, *268*, 177-183.
[http://dx.doi.org/10.1016/0003-2670(92)85262-5]

[61]   Divrikli, Ü.; Elçi, L. Determination of some trace metals in water and sediment samples by flame atomic absorption spectrometry after coprecipitation with cerium (IV) hydroxide. *Anal. Chim. Acta,* **2002**, *452*, 231-235.
[http://dx.doi.org/10.1016/S0003-2670(01)01462-3]

[62]   Agapito, R.; Alves, S.; Capelo, J.L.; Gonçalves, M.L.; Mota, A.M. Sample treatment with focused ultrasound and bath sonication as a powerful tool for the evaluation of cadmium pollution in estuarine waters. *Mar. Chem.,* **2006**, *98*, 286-294.
[http://dx.doi.org/10.1016/j.marchem.2005.10.006]

[63]   Varghese, A.; Khadar, A.M.A.; Kalluraya, B. Simultaneous determination of titanium and molybdenum in steel samples using derivative spectrophotometry in neutral micellar medium. *Spectrochim. Acta A Mol. Biomol. Spectrosc.,* **2006**, *64*(2), 383-390.
[http://dx.doi.org/10.1016/j.saa.2005.07.034] [PMID: 16384735]

[64]   El-Moselhy, K.M.; Gabal, M.N. Trace metals in water, sediments and marine organisms from the northern part of the Gulf of Suez, Red Sea. *J. Mar. Syst.,* **2004**, *46*, 39-46.
[http://dx.doi.org/10.1016/j.jmarsys.2003.11.014]

[65]   Haratake, M.; Yasumoto, K.; Ono, M.; Akashi, M.; Nakayama, M. Synthesis of hydrophilic macroporous chelating polymers and their versatility in the preconcentration of metals in seawater samples. *Anal. Chim. Acta,* **2006**, *561*, 183-190.
[http://dx.doi.org/10.1016/j.aca.2006.01.042]

[66]   Camel, V. Solid phase extraction of trace elements. *Spectrochim. Acta B At. Spectrosc.,* **2003**, *58*, 1177-1233.
[http://dx.doi.org/10.1016/S0584-8547(03)00072-7]

[67]   Mester, Z.; Sturgeon, R. Trace element speciation using solid phase microextraction. *Spectrochim. Acta B At. Spectrosc.,* **2005**, *60*, 1243-1269.
[http://dx.doi.org/10.1016/j.sab.2005.06.013]

[68]   Guibal, E. Interactions of metal ions with chitosan-based sorbents: a review. *Separ. Purif. Tech.,* **2004**, *38*, 43-74.
[http://dx.doi.org/10.1016/j.seppur.2003.10.004]

[69]   Hashemi, P.; Noresson, B.; Olin, A. Properties of a high capacity iminodiacetate-agarose adsorbent and its application in a flow system with on-line buffering of acidified samples for accumulation of metal ions in natural waters. *Talanta,* **1999**, *49*(4), 825-835.
[http://dx.doi.org/10.1016/S0039-9140(99)00079-X] [PMID: 18967658]

[70]   Wu, X.Z.; Liu, P.; Pu, Q.S.; Sun, Q.Y.; Su, Z.X. Preparation of dendrimer-like polyamidoamine immobilized silica gel and its application to online preconcentration and separation palladium prior to FAAS determination. *Talanta,* **2004**, *62*(5), 918-923.
[http://dx.doi.org/10.1016/j.talanta.2003.10.011] [PMID: 18969380]

[71]   Hashemi, P.; Boroumand, J.; Fat'hi, M.R. A dual column system using agarose-based adsorbents for preconcentration and speciation of chromium in water. *Talanta,* **2004**, *64*(3), 578-583.
[http://dx.doi.org/10.1016/j.talanta.2004.03.035] [PMID: 18969644]

[72]   Zhang, A.; Asakura, T.; Uchiyama, G. The adsorption mechanism of uranium(VI) from seawater on a macroporous fibrous polymeric adsorbent containing amidoxime chelating functional group. *React. Funct. Polym.,* **2003**, *57*, 67-76.
[http://dx.doi.org/10.1016/j.reactfunctpolym.2003.07.005]

[73]   Karolev, A. Substitutional complexometric determination of magnesium in homogeneous water-dioxan medium, based on decomposition of MgEDTA with 8-quinolinol and titration of released EDTA with calcium. *Talanta,* **1992**, *39*(12), 1575-1577.
[http://dx.doi.org/10.1016/0039-9140(92)80185-G] [PMID: 18965572]

[74]   Chmielarz, A.; Gnot, W. Conversion of zinc chloride to zinc sulphate by electrodialysis—a new

concept for solving the chloride ion problem in zinc hydrometallurgy. *Hydrometallurgy,* **2001**, *61*, 21-43.
[http://dx.doi.org/10.1016/S0304-386X(01)00153-0]

[75]   Lal, S. Control and chemical analysis of plating solutions. *Met. Finish.,* **2009**, *107*, 14-22.
[http://dx.doi.org/10.1016/S0026-0576(09)00043-9]

[76]   Rapti-Caputo, D.; Vaccaro, C. Geochemical evidences of landfill leachate in groundwater. *Eng. Geol.,*
2006, *85*, 111-121.
[http://dx.doi.org/10.1016/j.enggeo.2005.09.032]

[77]   Florence, T.M.; Batley, G.E. Determination of the chemical forms of trace metals in natural waters,
with special reference to copper, lead, cadmium and zinc. *Talanta,* **1977**, *24*(3), 151-158.
[http://dx.doi.org/10.1016/0039-9140(77)80080-5] [PMID: 18962054]

[78]   Itoh, J.; Liu, J.; Komata, M. Novel analytical applications of porphyrin to HPLC post-column flow
injection system for determination of the lanthanides. *Talanta,* **2006**, *69*(1), 61-67.
[http://dx.doi.org/10.1016/j.talanta.2005.08.056] [PMID: 18970532]

[79]   Greenhalgh, R.; Riley, J.P.; Tongudai, M. An ion-exchange scheme for the determination of the major
cations in sea water. *Anal. Chim. Acta,* **1966**, *36*, 439-448.
[http://dx.doi.org/10.1016/0003-2670(66)80078-8]

[80]   van Staden, J.F.; Taljaard, R.E. Determination of calcium in water, urine and pharmaceutical samples
by sequential injection analysis. *Anal. Chim. Acta,* **1996**, *323*, 75-85.
[http://dx.doi.org/10.1016/0003-2670(95)00615-X]

[81]   Garcia, C.A.B.; Júnior, L.R.; Neto, Gde.O. Determination of potassium ions in pharmaceutical
samples by FIA using a potentiometric electrode based on ionophore nonactin occluded in EVA
membrane. *J. Pharm. Biomed. Anal.,* **2003**, *31*(1), 11-18.
[http://dx.doi.org/10.1016/S0731-7085(02)00598-8] [PMID: 12560044]

[82]   Chen, M-J.; Hsieh, Y-T.; Weng, Y-M.; Chiou, R.Y-Y. Flame photometric determination of salinity in
processed foods. *Food Chem.,* **2005**, *91*, 765-770.
[http://dx.doi.org/10.1016/j.foodchem.2004.10.002]

[83]   Hernández, O.; Jiménez, F.; Jiménez, A.I.; Arias, J.J.; Havel, J. Multicomponent flow injection based
analysis with diode array detection and partial least squares multivariate calibration evaluation. Rapid
determination of Ca(II) and Mg(II) in waters and dialysis liquids. *Anal. Chim. Acta,* **1996**, *320*, 177-
183.
[http://dx.doi.org/10.1016/0003-2670(95)00528-5]

[84]   Cerdà, V.; Cerdà, A.; Cladera, A.; Oms, M.; Mas, F.; Gómez, E. Monitoring of environmental
parameters by sequential injection analysis. *TrAC. Trends Analyt. Chem.,* **2001**, *20*, 407-418.
[http://dx.doi.org/10.1016/S0165-9936(01)00064-4]

[85]   Poboży, E.; Halko, R.; Krasowski, M.; Wierzbicki, T.; Trojanowicz, M. Flow-injection sample
preconcentration for ion-pair chromatography of trace metals in waters. *Water Res.,* **2003**, *37*(9),
2019-2026.
[http://dx.doi.org/10.1016/S0043-1354(02)00615-2] [PMID: 12691886]

[86]   Bian, Q.Z.; Jacob, P.; Berndt, H.; Niemax, K. Online flow digestion of biological and environmental
samples for inductively coupled plasma–optical emission spectroscopy (ICP–OES). *Anal. Chim. Acta,*
2005, *538*, 323-329.
[http://dx.doi.org/10.1016/j.aca.2005.01.074]

[87]   van Staden, J.F.; Kluever, L.G. Determination of manganese in natural water and effluent streams
using a solid-phase lead(IV) dioxide reactor in a flow-injection system. *Anal. Chim. Acta,* **1997**, *350*,
15-20.
[http://dx.doi.org/10.1016/S0003-2670(97)00293-6]

[88]   Miró, M.; Estela, J.M.; Cerdà, V. Application of flowing stream techniques to water analysis Part III.

Metal ions: alkaline and alkaline-earth metals, elemental and harmful transition metals, and multielemental analysis. *Talanta,* **2004**, *63*(2), 201-223.
[PMID: 18969420]

[89]   Maity, S.; Chakravarty, S.; Thakur, P.; Gupta, K.K.; Bhattacharjee, S.; Roy, B.C. Evaluation and standardisation of a simple HG-AAS method for rapid speciation of As(III) and As(V) in some contaminated groundwater samples of West Bengal, India. *Chemosphere,* **2004**, *54*(8), 1199-1206.
[http://dx.doi.org/10.1016/j.chemosphere.2003.09.035] [PMID: 14664849]

[90]   Pehkonen, S. Determination of the oxidation states of iron in natural waters. A review. *Analyst (Lond.),* **1995**, *120*, 2655.
[http://dx.doi.org/10.1039/an9952002655]

[91]   Themelis, D.G.; Tzanavaras, P.D.; Kika, F.S.; Sofoniou, M.C. Flow-injection manifold for the simultaneous spectrophotometric determination of Fe(II) and Fe(III) using 2,2'-dipyridyl-2-pyridylhydrazone and a single-line double injection approach. *Fresenius J. Anal. Chem.,* **2001**, *371*(3), 364-368.
[http://dx.doi.org/10.1007/s002160100930] [PMID: 11688651]

[92]   Safavi, A.; Abdollahi, H.; Mirzajani, R. Simultaneous spectrophotometric determination of Fe(III), Al(III) and Cu(II) by partial least-squares calibration method. *Spectrochim. Acta A Mol. Biomol. Spectrosc.,* **2006**, *63*(1), 196-199.
[http://dx.doi.org/10.1016/j.saa.2005.05.004] [PMID: 15975847]

[93]   de Jesus, D.S.; Cassella, R.J.; Ferreira, S.L.C.; Costa, A.C.S.; de Carvalho, M.S.; Santelli, R.E. Polyurethane foam as a sorbent for continuous flow analysis: Preconcentration and spectrophotometric determination of zinc in biological materials. *Anal. Chim. Acta,* **1998**, *366*, 263-269.
[http://dx.doi.org/10.1016/S0003-2670(98)00111-1]

[94]   Ghasemi, J.; Ahmadi, S.; Torkestani, K. Simultaneous determination of copper, nickel, cobalt and zinc using zincon as a metallochromic indicator with partial least squares. *Anal. Chim. Acta,* **2003**, *487*, 181-188.
[http://dx.doi.org/10.1016/S0003-2670(03)00556-7]

[95]   van Staden, J.K.; Tlowana, S.S. On-line separation, simultaneous dilution and spectrophotometric determination of zinc in fertilisers with a sequential injection system and xylenol orange as complexing agent. *Talanta,* **2002**, *58*(6), 1115-1122.
[http://dx.doi.org/10.1016/S0039-9140(02)00409-5] [PMID: 18968847]

[96]   Asan, A.; Isildak, I.; Andac, M.; Yilmaz, F. A simple and selective flow-injection spectrophotometric determination of copper(II) by using acetylsalicylhydroxamic acid. *Talanta,* **2003**, *60*(4), 861-866.
[http://dx.doi.org/10.1016/S0039-9140(03)00134-6] [PMID: 18969111]

[97]   Rumori, P.; Cerdà, V. Reversed flow injection and sandwich sequential injection methods for the spectrophotometric determination of copper(II) with cuprizone. *Anal. Chim. Acta,* **2003**, *486*, 227-235.
[http://dx.doi.org/10.1016/S0003-2670(03)00493-8]

[98]   Gao, L.; Ren, S. Simultaneous spectrophotometric determination of four metals by two kinds of partial least squares methods. *Spectrochim. Acta A Mol. Biomol. Spectrosc.,* **2005**, *61*(13-14), 3013-3019.
[http://dx.doi.org/10.1016/j.saa.2004.11.020] [PMID: 16165045]

[99]   Woo Kang, S.; Saki, T.; Ohno, N.; Ida, K. Simultaneous spectrophotometric determination of iron and copper in serum with 2-(5-bromo-2-pyridylazo)-5-(N-propyl-N-sulphopropylamino) aniline by flow-injection analysis. *Anal. Chim. Acta,* **1992**, *261*, 197-203.
[http://dx.doi.org/10.1016/0003-2670(92)80191-9]

[100] Kolotyrkina, I.Y.; Shpigun, L.K.; Zolotov, Y.A.; Tsysin, G.I. Shipboard flow injection method for the determination of manganese in sea-water using in-valve preconcentration and catalytic spectrophotometric detection. *Analyst (Lond.),* **1991**, *116*, 707.
[http://dx.doi.org/10.1039/an9911600707]

[101] Hirayama, K.; Unohara, N. Spectrophotometric catalytic determination of an ultratrace amount of

iron(III) in water based on the oxidation of N,N-dimethyl-p-phenylenediamine by hydrogen peroxide. *Anal. Chem.,* **1988**, *60*, 2573-2577.
[http://dx.doi.org/10.1021/ac00174a009]

[102]　Kolotyrkina, I.Y.; Shpigun, L.K.; Zolotov, Y.A.; Malahoff, A. Application of flow injection spectrophotometry to the determination of dissolved iron in sea-water. *Analyst (Lond.),* **1995**, *120*, 201.
[http://dx.doi.org/10.1039/an9952000201]

[103]　Ojeda, C.B.; Rojas, F.S. Determination of rhodium: Since the origins until today Spectrophotometric methods. *Talanta,* **2005**, *67*(1), 1-19.
[http://dx.doi.org/10.1016/j.talanta.2005.02.028] [PMID: 18970131]

[104]　Prasad, S. Kinetic method for determination of nanogram amounts of copper(II) by its catalytic effect on hexacynoferrate(III)–citric acid indicator reaction. *Anal. Chim. Acta,* **2005**, *540*, 173-180.
[http://dx.doi.org/10.1016/j.aca.2005.03.011]

[105]　Fang, Z.; Guo, T.; Welz, B. Determination of cadmium, lead and copper in water samples by flame atomic-absorption spectrometry with preconcentration by flow-injection on-line sorbent extraction. *Talanta,* **1991**, *38*(6), 613-619.
[http://dx.doi.org/10.1016/0039-9140(91)80144-O] [PMID: 18965193]

[106]　Olsen, S.; Pessenda, L.C.R.; Růžička, J.; Hansen, E.H. Combination of flow injection analysis with flame atomic-absorption spectrophotometry: determination of trace amounts of heavy metals in polluted seawater. *Analyst (Lond.),* **1983**, *108*, 905-917.
[http://dx.doi.org/10.1039/AN9830800905]

[107]　Croot, P.L.; Laan, P. Continuous shipboard determination of Fe(II) in polar waters using flow injection analysis with chemiluminescence detection. *Anal. Chim. Acta,* **2002**, *466*, 261-273.
[http://dx.doi.org/10.1016/S0003-2670(02)00596-2]

[108]　Vink, S.; Boyle, E.; Measures, C.; Yuan, J. Automated high resolution determination of the trace elements iron and aluminium in the surface ocean using a towed Fish coupled to flow injection analysis. *Deep Sea Res. Part I Oceanogr. Res. Pap.,* **2000**, *47*, 1141-1156.
[http://dx.doi.org/10.1016/S0967-0637(99)00074-6]

[109]　Lannuzel, D.; de Jong, J.; Schoemann, V.; Trevena, A.; Tison, J-L.; Chou, L. Development of a sampling and flow injection analysis technique for iron determination in the sea ice environment. *Anal. Chim. Acta,* **2006**, *556*, 476-483.
[http://dx.doi.org/10.1016/j.aca.2005.09.059]

[110]　Flame Atomic Absorption Spectrometry. **2020**. Available from: http://chemicalinstrumen tation.weebly.com/flame-aas.html

[111]　Skoog, D.A.; Holler, F.J.; Crouch, S.R. *Principles of instrumental analysis,* 7th ed; Thomson Brooks/Cole, **2018**.

[112]　Roldan, P.S.; Alcântara, I.L.; Castro, G.R.; Rocha, J.C.; Padilha, C.C.F.; Padilha, P.M. Determination of Cu, Ni, and Zn in fuel ethanol by FAAS after enrichment in column packed with 2-aminothiazol--modified silica gel. *Anal. Bioanal. Chem.,* **2003**, *375*(4), 574-577.
[http://dx.doi.org/10.1007/s00216-002-1735-7] [PMID: 12610713]

[113]　da Silva, E.L.; Ganzarolli, E.M.; Carasek, E. Use of Nb_2O_5-SiO_2 in an automated on-line preconcentration system for determination of copper and cadmium by FAAS. *Talanta,* **2004**, *62*(4), 727-733.
[http://dx.doi.org/10.1016/j.talanta.2003.09.014] [PMID: 18969355]

[114]　Araujo, P.W.; Gómez, M.J.; de Benzo, Z.A.; Castillo, C. Use of experimental designs for estimation of optimum electrothermal atomic absorption conditions for molybdenum. *Chemom. Intell. Lab. Syst.,* **1992**, *16*, 203-211.
[http://dx.doi.org/10.1016/0169-7439(92)80038-6]

[115] Matthews, D.O.; McGahan, M.C. Cysteine—an effective matrix modifier for determination of gold in biological samples by electrothermal atomic absorption spectrophotometry. *Spectrochim. Acta B At. Spectrosc.,* **1987**, *42*, 909-913.
[http://dx.doi.org/10.1016/0584-8547(87)80101-5]

[116] Bragigand, V.; Berthet, B.; Amiard, J.C.; Rainbow, P.S. Estimates of trace metal bioavailability to humans ingesting contaminated oysters. *Food Chem. Toxicol.,* **2004**, *42*(11), 1893-1902.
[http://dx.doi.org/10.1016/j.fct.2004.07.011] [PMID: 15350688]

[117] Falomir, P.; Alegría, A.; Barberá, R.; Farré, R.; Lagarda, M.J. Direct determination of lead in human milk by electrothermal atomic absorption spectrometry. *Food Chem.,* **1999**, *64*, 111-113.
[http://dx.doi.org/10.1016/S0308-8146(98)00116-2]

[118] Ampan, P.; Ruzicka, J.; Atallah, R.; Christian, G.D.; Jakmunee, J.; Grudpan, K. Exploiting sequential injection analysis with bead injection and lab-on-valve for determination of lead using electrothermal atomic absorption spectrometry. *Anal. Chim. Acta,* **2003**, *499*, 167-172.
[http://dx.doi.org/10.1016/j.aca.2003.08.030]

[119] Pedro, J.; Stripekis, J.; Bonivardi, A.; Tudino, M. Thermal stabilization of tellurium in mineral acids solutions: use of permanent modifiers for its determination in sulfur by GFAAS. *Talanta,* **2006**, *69*(1), 199-203.
[http://dx.doi.org/10.1016/j.talanta.2005.09.027] [PMID: 18970554]

[120] Carrión, N.; Itriago, A.M.; Alvarez, M.A.; Eljuri, E. Simultaneous determination of lead, nickel, tin and copper in aluminium-base alloys using slurry sampling by electrical discharge and multielement ETAAS. *Talanta,* **2003**, *61*(5), 621-632.
[http://dx.doi.org/10.1016/S0039-9140(03)00363-1] [PMID: 18969226]

[121] Mofolo, R.M.; Katskov, D.A.; Tittarelli, P.; Grotti, M. Vaporization of indium nitrate in the graphite tube atomizer in the presence of chemical modifiers. *Spectrochim. Acta B At. Spectrosc.,* **2001**, *56*, 375-391.
[http://dx.doi.org/10.1016/S0584-8547(01)00167-7]

[122] Chan, M.S.; Huang, S.D. Direct determination of cadmium and copper in seawater using a transversely heated graphite furnace atomic absorption spectrometer with Zeeman-effect background corrector. *Talanta,* **2000**, *51*(2), 373-380.
[http://dx.doi.org/10.1016/S0039-9140(99)00283-0] [PMID: 18967869]

[123] Xiao-Quan, S.; Radziuk, B.; Welz, B.; Vyskočilová, O. Determination of manganese in river and sea-water samples by electrothermal atomic absorption spectrometry with a tungsten atomizer. *J. Anal. At. Spectrom.,* **1993**, *8*, 409-413.
[http://dx.doi.org/10.1039/JA9930800409]

[124] Rust, J.A.; Nóbrega, J.A.; Calloway, C.P.; Jones, B.T. Advances with tungsten coil atomizers: Continuum source atomic absorption and emission spectrometry. S*pectrochim. Acta - Part B At. Spectroscopy (Springf.),* **2005**, *60*, 589-598.

[125] Nóbrega, J.A.; Rust, J.; Calloway, C.P.; Jones, B.T. Use of modifiers with metal atomizers in electrothermal atomic absorption spectrometry: A short review. *Spectrochim. Acta B At. Spectrosc.,* **2004**, *59*, 1337-1345.
[http://dx.doi.org/10.1016/j.sab.2004.06.003]

[126] Godlewska-Zyłkiewicz, B.; Zaleska, M. Preconcentration of palladium in a flow-through electrochemical cell for determination by graphite furnace atomic absorption spectrometry. *Anal. Chim. Acta,* **2002**, *462*, 305-312.
[http://dx.doi.org/10.1016/S0003-2670(02)00353-7]

[127] Hou, X.; Jones, B.T. Tungsten devices in analytical atomic spectrometry. *Spectrochim. Acta B At. Spectrosc.,* **2002**, *57*, 659-688.
[http://dx.doi.org/10.1016/S0584-8547(02)00014-9]

[128] Zhang, A.; Uchiyama, G.; Asakura, T. pH Effect on the uranium adsorption from seawater by a macroporous fibrous polymeric material containing amidoxime chelating functional group. *React. Funct. Polym.,* **2005**, *63*, 143-153.
[http://dx.doi.org/10.1016/j.reactfunctpolym.2005.02.015]

[129] Navarro, R.R.; Wada, S.; Tatsumi, K. Heavy metal precipitation by polycation-polyanion complex of PEI and its phosphonomethylated derivative. *J. Hazard. Mater.,* **2005**, *123*(1-3), 203-209.
[http://dx.doi.org/10.1016/j.jhazmat.2005.03.048] [PMID: 15925445]

[130] Sturgeon, R.E.; Berman, S.S.; Desaulniers, A.; Russell, D.S. Pre-concentration of trace metals from sea-water for determination by graphite-furnace atomic-absorption spectrometry. *Talanta,* **1980**, *27*(2), 85-94.
[http://dx.doi.org/10.1016/0039-9140(80)80023-3] [PMID: 18962623]

[131] Cabon, J.Y.; Le Bihan, A. Direct determination of zinc in sea-water using electrothermal atomic absorption spectrometry with Zeeman-effect background correction: Effects of chemical and spectral interferences. *J. Anal. At. Spectrom.,* **1994**, *9*, 477-481.
[http://dx.doi.org/10.1039/ja9940900477]

[132] Som-Aum, W.; Liawruangrath, S.; Hansen, E.H. Flow injection on-line preconcentration of low levels of Cr(VI) with detection by ETAAS: Comparison of using an open tubular PTFE knotted reactor and a column reactor packed with PTFE beads. *Anal. Chim. Acta,* **2002**, *463*, 99-109.
[http://dx.doi.org/10.1016/S0003-2670(02)00349-5]

[133] Wang, J.; Hansen, E.H. Coupling on-line preconcentration by ion-exchange with ETAAS: A novel flow injection approach based on the use of a renewable microcolumn as demonstrated for the determination of nickel in environmental and biological samples. *Anal. Chim. Acta,* **2000**, *424*, 223-232.
[http://dx.doi.org/10.1016/S0003-2670(00)01083-7]

[134] Sung, Y.H.; Liu, Z.S.; Huang, S.D. Automated on-line preconcentration system for electrothermal atomic absorption spectrometry for the determination of copper and molybdenum in sea-water. *J. Anal. At. Spectrom.,* **1997**, *12*, 841-847.
[http://dx.doi.org/10.1039/a701657c]

[135] Wang, J.; Harald Hansen, E. Trends and perspectives of flow injection/sequential injection on-line sample-pretreatment schemes coupled to ETAAS. *TrAC -. Trends Analyt. Chem.,* **2005**, *24*, 1-8.
[http://dx.doi.org/10.1016/j.trac.2004.08.011]

[136] Huang, J.; Hu, X.; Zhang, J.; Li, K.; Yan, Y.; Xu, X. The application of inductively coupled plasma mass spectrometry in pharmaceutical and biomedical analysis. *J. Pharm. Biomed. Anal.,* **2006**, *40*(2), 227-234.
[http://dx.doi.org/10.1016/j.jpba.2005.11.014] [PMID: 16364586]

[137] Rosen, A.L.; Hieftje, G.M. Inductively coupled plasma mass spectrometry and electrospray mass spectrometry for speciation analysis: Applications and instrumentation. *Spectrochim. Acta B At. Spectrosc.,* **2004**, *59*, 135-146.
[http://dx.doi.org/10.1016/j.sab.2003.09.004]

[138] Lee, J.; Park, J. Il.; Youn, Y.S.; Ha, Y.K.; Kim, J.Y. Application of Laser Ablation Inductively Coupled Plasma Mass Spectrometry for Characterization of U-7Mo/Al-5Si Dispersion Fuels. *Nucl. Eng. Technol.,* **2017**, *49*, 645-650.
[http://dx.doi.org/10.1016/j.net.2016.08.014]

[139] Holliday, A.E.; Beauchemin, D. Spatial profiling of analyte signal intensities in inductively coupled plasma mass spectrometry. *Spectrochim. Acta B At. Spectrosc.,* **2004**, *59*, 291-311.
[http://dx.doi.org/10.1016/j.sab.2003.12.018]

[140] Gäckle, M.; Merten, D. Modelling the temporal intensity distribution in laser ablation-inductively coupled plasma-mass spectrometry (LA-ICP-MS) using scanning and drilling mode. *Spectrochim. Acta B At. Spectrosc.,* **2005**, *60*, 1517-1530.

[http://dx.doi.org/10.1016/j.sab.2005.10.001]

[141]   Krachler, M.; Zheng, J.; Fisher, D.; Shotyk, W. Analytical procedures for improved trace element detection limits in polar ice from Arctic Canada using ICP-SMS. *Anal. Chim. Acta,* **2005**, *530*, 291-298.
[http://dx.doi.org/10.1016/j.aca.2004.09.024]

[142]   Velitchkova, N.; Pentcheva, E.N.; Daskalova, N. Determination of arsenic, mercury, selenium, thallium, tin and bismuth in environmental materials by inductively coupled plasma emission spectrometry. *Spectrochim. Acta B At. Spectrosc.,* **2004**, *59*, 871-882.
[http://dx.doi.org/10.1016/j.sab.2004.03.004]

[143]   Water for ICP-MS. Available from: https://www.merckmillipore.com/INTL/en/water-purification/learning-centers/applications/inorganic-analysis/icp-ms/_e2b.qB.s7QAAAFAniQQWTtN, nav?ReferrerURL=https%3A%2F%2Fwww.google.com%2F

[144]   Bettinelli, M.; Spezia, S. Determination of trace elements in sea water by ion chromatography-inductively coupled plasma mass spectrometry. *J. Chromatogr. A,* **1995**, *709*, 275-281.
[http://dx.doi.org/10.1016/0021-9673(95)00454-U]

[145]   Ramesh, A.; Rama Mohan, K.; Seshaiah, K. Preconcentration of trace metals on Amberlite XAD-4 resin coated with dithiocarbamates and determination by inductively coupled plasma-atomic emission spectrometry in saline matrices. *Talanta,* **2002**, *57*(2), 243-252.
[http://dx.doi.org/10.1016/S0039-9140(02)00033-4] [PMID: 18968624]

[146]   Benkhedda, K.; Infante, H.G.; Adams, F.C.; Ivanova, E. Inductively coupled plasma mass spectrometry for trace analysis using flow injection on-line preconcentration and time-of-flight mass analyser. *TrAC -. Trends Analyt. Chem.,* **2002**, *21*, 332-342.
[http://dx.doi.org/10.1016/S0165-9936(02)00501-0]

[147]   Wang, W.; Fthenakis, V. Kinetics study on separation of cadmium from tellurium in acidic solution media using ion-exchange resins. *J. Hazard. Mater.,* **2005**, *125*(1-3), 80-88.
[http://dx.doi.org/10.1016/j.jhazmat.2005.02.013] [PMID: 16118035]

[148]   Tel, H.; Altaş, Y.; Taner, M.S. Adsorption characteristics and separation of Cr(III) and Cr(VI) on hydrous titanium(IV) oxide. *J. Hazard. Mater.,* **2004**, *112*(3), 225-231.
[http://dx.doi.org/10.1016/j.jhazmat.2004.05.025] [PMID: 15302443]

[149]   Coedo, A.G.; Dorado, M.T.; Padilla, I.; Alguacil, F.J. Study of the application of air-water flow injection inductively coupled plasma mass spectrometry for the determination of calcium in steels. *J. Anal. At. Spectrom.,* **1996**, *11*, 1037-1041.
[http://dx.doi.org/10.1039/ja9961101037]

[150]   Benkhedda, K.; Infante, H.G.; Adams, F.C. Determination of total lead and lead isotope ratios in natural waters by inductively coupled plasma time-of-flight mass spectrometry after flow injection on-line preconcentration. *Anal. Chim. Acta,* **2004**, *506*, 137-144.
[http://dx.doi.org/10.1016/j.aca.2003.11.007]

[151]   Di Nezio, M.S.; Palomeque, M.E.; Fernández Band, B.S. A sensitive spectrophotometric method for lead determination by flow injection analysis with on-line preconcentration. *Talanta,* **2004**, *63*(2), 405-409.
[http://dx.doi.org/10.1016/j.talanta.2003.11.012] [PMID: 18969447]

[152]   Hirata, S.; Ishida, Y.; Aihara, M.; Honda, K.; Shikino, O. Determination of trace metals in seawater by on-line column preconcentration inductively coupled plasma mass spectrometry. *Anal. Chim. Acta,* **2001**, *438*, 205-214.
[http://dx.doi.org/10.1016/S0003-2670(01)00859-5]

[153]   Broekaert, J.A.C.; Leis, F. An injection method for the sequential determination of boron and several metals in waste-water samples by inductively-coupled plasma atomic emission spectrometry. *Anal. Chim. Acta,* **1979**, *109*, 73-83.
[http://dx.doi.org/10.1016/S0003-2670(01)84229-X]

[154] Pickhardt, C.; Izmer, A.V.; Zoriy, M.V.; Schaumlöffel, D.; Sabine Becker, J. On-line isotope dilution in laser ablation inductively coupled plasma mass spectrometry using a microflow nebulizer inserted in the laser ablation chamber. *Int. J. Mass Spectrom.,* **2006**, *248*, 136-141.
[http://dx.doi.org/10.1016/j.ijms.2005.11.001]

[155] Huang, Z.Y.; Chen, F.R.; Zhuang, Z.X.; Wang, X.R.; Lee, F.S.C. Trace lead measurement and on-line removal of matrix interference in seawater by isotope dilution coupled with flow injection and ICP-MS. *Anal. Chim. Acta,* **2004**, *508*, 239-245.
[http://dx.doi.org/10.1016/j.aca.2003.11.033]

[156] Liu, H.W.; Jiang, S.J.; Liu, S.H. Determination of cadmium, mercury and lead in seawater by electrothermal vaporization isotope dilution inductively coupled plasma mass spectrometry. *Spectrochim. Acta B At. Spectrosc.,* **1999**, *54*, 1367-1375.
[http://dx.doi.org/10.1016/S0584-8547(99)00081-6]

[157] Filipović-Kovacević, Z.; Sipos, L. Voltammetric determination of copper in water samples digested by ozone. *Talanta,* **1998**, *45*(5), 843-850.
[http://dx.doi.org/10.1016/S0039-9140(97)00174-4] [PMID: 18967069]

[158] Miró, M.; Estela, J.M.; Cerdà, V. Application of flowing stream techniques to water analysis Part III. Metal ions: alkaline and alkaline-earth metals, elemental and harmful transition metals, and multielemental analysis. *Talanta,* **2004**, *63*(2), 201-223.
[PMID: 18969420]

[159] Ellwood, M.J.; Van Den Berg, C.M.G. Zinc speciation in the Northeastern Atlantic Ocean. *Mar. Chem.,* **2000**, *68*, 295-306.
[http://dx.doi.org/10.1016/S0304-4203(99)00085-7]

[160] Muller, F.L.L.; Kester, D.R. Voltammetric determination of the complexation parameters of zinc in marine and estuarine waters. *Mar. Chem.,* **1991**, *33*, 71-90.
[http://dx.doi.org/10.1016/0304-4203(91)90058-5]

[161] Chow, C.W.K.; Kolev, S.D.; Davey, D.E.; Mulcahy, D.E. Determination of copper in natural waters by batch and oscillating flow injection stripping potentiometry. *Anal. Chim. Acta,* **1996**, *330*, 79-87.
[http://dx.doi.org/10.1016/0003-2670(96)00160-2]

[162] Turner, A.; Nimmo, M.; Thuresson, K.A. Speciation and sorptive behaviour of nickel in an organic-rich estuary (Beaulieu, UK). *Mar. Chem.,* **1998**, *63*, 105-118.
[http://dx.doi.org/10.1016/S0304-4203(98)00054-1]

[163] Kiptoo, J.K.; Ngila, J.C.; Sawula, G.M. Speciation studies of nickel and chromium in wastewater from an electroplating plant. *Talanta,* **2004**, *64*(1), 54-59.
[http://dx.doi.org/10.1016/j.talanta.2004.03.020] [PMID: 18969568]

[164] Tonle, I.K.; Ngameni, E.; Walcarius, A. Preconcentration and voltammetric analysis of mercury(II) at a carbon paste electrode modified with natural smectite-type clays grafted with organic chelating groups. *Sens. Actuators B Chem.,* **2005**, *110*, 195-203.
[http://dx.doi.org/10.1016/j.snb.2005.01.027]

[165] Nürnberg, H.W. Trace analytical procedures with modern voltammetric determination methods for the investigation and monitoring of ecotoxic heavy metals in natural waters and atmospheric precipitates. *Sci. Total Environ.,* **1984**, *37*, 9-34.
[http://dx.doi.org/10.1016/0048-9697(84)90113-X]

[166] Labuda, J.; Bučková, M. Selectivity of voltammetric determination at an ion-exchanger modified electrode. *Electrochem. Commun.,* **2000**, *2*, 322-324.
[http://dx.doi.org/10.1016/S1388-2481(00)00030-8]

[167] Alonso, E.; Santos, A.; Callejón, M.; Jiménez, J.C. Speciation as a screening tool for the determination of heavy metal surface water pollution in the Guadiamar river basin. *Chemosphere,* **2004**, *56*(6), 561-570.

[http://dx.doi.org/10.1016/j.chemosphere.2004.04.031] [PMID: 15212899]

[168] Van Staden, J.F.; Matoetoe, M.C. Simultaneous determination of copper, lead, cadmium and zinc using differential pulse anodic stripping voltammetry in a flow system. *Anal. Chim. Acta,* **2000**, *411,* 201-207.
[http://dx.doi.org/10.1016/S0003-2670(00)00785-6]

[169] Omanović, D.; Branica, M. Pseudopolarography of trace metals: Part I. The automatic ASV measurements of reversible electrode reactions. *J. Electroanal. Chem. (Lausanne),* **2003**, *543,* 83-92.
[http://dx.doi.org/10.1016/S0022-0728(02)01484-5]

[170] Daniele, S.; Bragato, C.; Baldo, M.A. An approach to the calibrationless determination of copper and lead by anodic stripping voltammetry at thin mercury film microelectrodes. Application to well water and rain. *Anal. Chim. Acta,* **1997**, *346,* 145-156.
[http://dx.doi.org/10.1016/S0003-2670(97)00114-1]

[171] Donat, J.R.; Lao, K.A.; Bruland, K.W. Speciation of dissolved copper and nickel in South San Francisco Bay: a multi-method approach. *Anal. Chim. Acta,* **1994**, *284,* 547-571.
[http://dx.doi.org/10.1016/0003-2670(94)85061-5]

[172] Murimboh, J.; Lam, M.T.; Hassan, N.M.; Chakrabarti, C.L. A study of Nafion-coated and uncoated thin mercury film-rotating disk electrodes for cadmium and lead speciation in model solutions of fulvic acid. *Anal. Chim. Acta,* **2000**, *423,* 115-126.
[http://dx.doi.org/10.1016/S0003-2670(00)01075-8]

[173] Honeychurch, K.C.; Hart, J.P. Screen-printed electrochemical sensors for monitoring metal pollutants. *TrAC -. Trends Analyt. Chem.,* **2003**, *22,* 456-469.
[http://dx.doi.org/10.1016/S0165-9936(03)00703-9]

[174] Hurst, M.P.; Bruland, K.W. The use of Nafion-coated thin mercury film electrodes for the determination of the dissolved copper speciation in estuarine water. *Anal. Chim. Acta,* **2005**, *546*(1), 68-78.
[http://dx.doi.org/10.1016/j.aca.2005.05.015] [PMID: 29569557]

[175] Kim, H.J.; Yun, K.S.; Yoon, E.; Kwak, J. A direct analysis of nanomolar metal ions in environmental water samples with Nafion-coated microelectrodes. *Electrochim. Acta,* **2004**, *50,* 205-210.
[http://dx.doi.org/10.1016/j.electacta.2004.07.031]

[176] Schöning, M.J.; Hüllenkremer, B.; Glück, O.; Lüth, H.; Emons, H. Voltohmmetry - A novel sensing principle for heavy metal determination in aqueous solutions. *Sens. Actuators B Chem.,* **2001**, *76,* 275-280.
[http://dx.doi.org/10.1016/S0925-4005(01)00582-2]

[177] Boye, M.; Aldrich, A.P.; Van Den Berg, C.M.G.; De Jong, J.T.M.; Veldhuis, M.; De Baar, H.J.W. Horizontal gradient of the chemical speciation of iron in surface waters of the northeast Atlantic Ocean. *Mar. Chem.,* **2003**, *80,* 129-143.
[http://dx.doi.org/10.1016/S0304-4203(02)00102-0]

[178] Bowie, A.R.; Whitworth, D.J.; Achterberg, E.P.; Mantoura, R.F.C.; Worsfold, P.J. Biogeochemistry of Fe and other trace elements (Al, Co, Ni) in the upper Atlantic Ocean. *Deep Res. Part I Oceanog.r Res. Pap.,* **2002**, *49,* 605-636.

[179] Trojánek, A.; Opekar, F. Flow-through coulometric stripping analysis and the determination of manganese by cathodic stripping voltammetry. *Anal. Chim. Acta,* **1981**, *126,* 15-21.
[http://dx.doi.org/10.1016/S0003-2670(01)83924-6]

[180] Roitz, J.S.; Bruland, K.W. Determination of dissolved manganese(II) in coastal and estuarine waters by differential pulse cathodic stripping voltammetry. *Anal. Chim. Acta,* **1997**, *344,* 175-180.
[http://dx.doi.org/10.1016/S0003-2670(97)00041-X]

[181] Knápek, J.; Komárek, J.; Krásenský, P. Determination of cadmium by electrothermal atomic absorption spectrometry using electrochemical separation in a microcell. *Spectrochim. Acta B At.*

Spectrosc., **2005**, *60*, 393-398.
[http://dx.doi.org/10.1016/j.sab.2004.12.011]

[182] Grotti, M.; Soggia, F.; Abelmoschi, M.L.; Rivaro, P.; Magi, E.; Frache, R. Temporal distribution of trace metals in Antarctic coastal waters. *Mar. Chem.,* **2001**, *76*, 189-209.
[http://dx.doi.org/10.1016/S0304-4203(01)00063-9]

[183] David Smith, J.; Redmond, J.D. Anodic stripping voltammetry applied to trace metals in seawater. *J. Electroanal. Chem.,* **1971**, *33*, 169-175.
[http://dx.doi.org/10.1016/S0022-0728(71)80218-8]

[184] O'Halloran, R.J. Anodic stripping voltammetry of manganese in seawater at a mercury film electrode. *Anal. Chim. Acta,* **1982**, *140*, 51-58.
[http://dx.doi.org/10.1016/S0003-2670(01)95451-0]

[185] van Dijck, G.; Verbeek, F. Determination of lead, cadmium, zinc and manganese in copper by anodic stripping voltammetry. *Anal. Chim. Acta,* **1971**, *54*, 475-481.
[http://dx.doi.org/10.1016/S0003-2670(01)82156-5]

[186] Abbas, M.N.; Zahran, E. Novel solid-state cadmium ion-selective electrodes based on its tetraiodo- and tetrabromo-ion pairs with cetylpyridinium. *J. Electroanal. Chem. (Lausanne),* **2005**, *576*, 205-213.
[http://dx.doi.org/10.1016/j.jelechem.2004.10.017]

[187] Cheng, K.L. Recent development of non-faradaic potentiometry. *Microchem. J.,* **2002**, *72*, 269-276.
[http://dx.doi.org/10.1016/S0026-265X(02)00092-9]

[188] Naja, G.; Mustin, C.; Volesky, B.; Berthelin, J. Stabilization of the initial electrochemical potential for a metal-based potentiometric titration study of a biosorption process. *Chemosphere,* **2006**, *62*(1), 163-170.
[http://dx.doi.org/10.1016/j.chemosphere.2005.03.055] [PMID: 16325652]

[189] Tymecki, Ł.; Zwierkowska, E.; Głąb, S.; Koncki, R. Strip thick-film silver ion-selective electrodes. *Sens. Actuators B Chem.,* **2003**, *96*, 482-488.
[http://dx.doi.org/10.1016/S0925-4005(03)00622-1]

[190] Li, Z.Q.; Wu, Z.Y.; Yuan, R.; Ying, M.; Shen, G.L.; Yu, R.Q. Thiocyanate-selective PVC membrane electrodes based on Mn(II) complex of N,N′-bis-(4-phenylazosalicylidene) o-phenylene diamine as a neutral carrier. *Electrochim. Acta,* **1999**, *44*, 2543-2548.
[http://dx.doi.org/10.1016/S0013-4686(98)00361-2]

[191] Singh, R.P.; Abbas, N.M.; Smesko, S.A. Suppressed ion chromatographic analysis of anions in environmental waters containing high salt concentrations. *J. Chromatogr. A,* **1996**, *733*, 73-91.
[http://dx.doi.org/10.1016/0021-9673(95)00957-4]

[192] Padarauskas, A.; Olšauskaite, V.; Paliulionyte, V. Simultaneous determination of inorganic anions and cations in waters by capillary electrophoresis. *J. Chromatogr. A,* **1998**, *829*(1-2), 359-365.
[http://dx.doi.org/10.1016/S0021-9673(98)00883-8] [PMID: 9923087]

[193] Ohta, K. Ion chromatography with indirect photometric detection of common mono- and divalent cations using an unmodified silica gel column and tyramine/oxalic acid/18-crown-6 as eluent. *Anal. Chim. Acta,* **2000**, *405*, 277-284.
[http://dx.doi.org/10.1016/S0003-2670(99)00696-0]

[194] Kubáň, P.; Guchardi, R.; Hauser, P.C. Trace-metal analysis with separation methods. *TrAC -. Trends Analyt. Chem.,* **2005**, *24*, 192-198.
[http://dx.doi.org/10.1016/j.trac.2004.11.012]

[195] Shaw, M.J.; Haddad, P.R. The determination of trace metal pollutants in environmental matrices using ion chromatography. *Environ. Int.,* **2004**, *30*(3), 403-431.
[http://dx.doi.org/10.1016/j.envint.2003.09.009] [PMID: 14987873]

[196] Ding, X.J.; Mou, S.F.; Liu, K.N.; Siriraks, A.; Riviello, J. Ion chromatography of heavy and transition metals by on- and post-column derivatizations. *Anal. Chim. Acta,* **2000**, *407*, 319-326.

[http://dx.doi.org/10.1016/S0003-2670(99)00798-9]

[197]   Mantoura, R.F.C.; Woodward, E.M.S. Conservative behaviour of riverine dissolved organic carbon in the Severn Estuary: chemical and geochemical implications. *Geochim. Cosmochim. Acta,* **1983**, *47*, 1293-1309.
[http://dx.doi.org/10.1016/0016-7037(83)90069-8]

[198]   Cruz-Morales, J.A.; Guadarrama, P. Synthesis, characterization and computational modeling of cyclen substituted with dendrimeric branches. Dendrimeric and macrocyclic moieties working together in a collective fashion. *J. Mol. Struct.,* **2005**, *779*, 1-10.
[http://dx.doi.org/10.1016/j.molstruc.2005.06.035]

[199]   Kim, J.K.; Kim, J.S.; Shul, Y.G.; Lee, K.W.; Oh, W.Z. Selective extraction of cesium ion with calix[4]arene crown ether through thin sheet supported liquid membranes. *J. Membr. Sci.,* **2001**, *187*, 3-11.
[http://dx.doi.org/10.1016/S0376-7388(00)00592-5]

[200]   Brinkman, U.A.T.; de Vries, G.; Kuroda, R. Thin-layer chromatographic data for inorganic substances. *J. Chromatogr. A,* **1973**, *85*, 187-520.
[http://dx.doi.org/10.1016/S0021-9673(01)83160-5]

[201]   Nageswara Rao, R.; Nagaraju, V. An overview of the recent trends in development of HPLC methods for determination of impurities in drugs. *J. Pharm. Biomed. Anal.,* **2003**, *33*(3), 335-377.
[http://dx.doi.org/10.1016/S0731-7085(03)00293-0] [PMID: 14550856]

[202]   Shi, Z.; Fu, C. Porphyrins as ligands for trace metal analysis by high-performance liquid chromatography. *Talanta,* **1997**, *44*(4), 593-604.
[http://dx.doi.org/10.1016/S0039-9140(96)02068-1] [PMID: 18966779]

[203]   Pobozy, E.; Głód, B.; Kaniewska, J.; Trojanowicz, M. Determination of triorganotin compounds by ion chromatography and capillary electrophoresis with preconcentration using solid-phase extraction. *J. Chromatogr. A,* **1995**, *718*, 329-338.
[http://dx.doi.org/10.1016/0021-9673(95)00696-6]

[204]   Molina, C.B.; Casas, J.A.; Zazo, J.A.; Rodríguez, J.J. A comparison of Al-Fe and Zr-Fe pillared clays for catalytic wet peroxide oxidation. *Chem. Eng. J.,* **2006**, *118*, 29-35.
[http://dx.doi.org/10.1016/j.cej.2006.01.007]

[205]   Nowack, B. Environmental chemistry of phosphonates. *Water Res.,* **2003**, *37*(11), 2533-2546.
[http://dx.doi.org/10.1016/S0043-1354(03)00079-4] [PMID: 12753831]

[206]   Mackey, D.J. Metal-organic complexes in seawater - An investigation of naturally occurring complexes of Cu, Zn, Fe, Mg, Ni, Cr, Mn and Cd using high-performance liquid chromatography with atomic fluorescence detection. *Mar. Chem.,* **1983**, *13*, 169-180.
[http://dx.doi.org/10.1016/0304-4203(83)90012-9]

[207]   Ruedas Rama, M.J.; Ruiz Medina, A.; Molina Díaz, A. A flow-injection renewable surface sensor for the fluorimetric determination of vanadium(V) with Alizarin Red S. *Talanta,* **2005**, *66*(5), 1333-1339.
[http://dx.doi.org/10.1016/j.talanta.2005.01.053] [PMID: 18970126]

[208]   De Armas, G.; Miró, M.; Cladera, A.; Estela, J.M.; Cerdà, V. Time-based multisyringe flow injection system for the spectrofluorimetric determination of aluminium. *Anal. Chim. Acta,* **2002**, *455*, 149-157.
[http://dx.doi.org/10.1016/S0003-2670(01)01588-4]

[209]   Price, D.; Worsfold, P.J.; Mantoura, R.F. Hydrogen peroxide in the marine environment: cycling and methods of analysis. *Trends Analyt. Chem.,* **1992**, *11*, 379-384.
[http://dx.doi.org/10.1016/0165-9936(92)80028-5]

[210]   Abou-Mesalam, M.M. Sorption kinetics of copper, zinc, cadmium and nickel ions on synthesized silico-antimonate ion exchanger. *Colloids Surf. A Physicochem. Eng. Asp.,* **2003**, *225*, 85-94.
[http://dx.doi.org/10.1016/S0927-7757(03)00191-2]

[211]   De Vito, I.E.; Masi, A.N.; Olsina, R.A. Determination of trace rare earth elements by X-ray

fluorescence spectrometry after preconcentration on a new chelating resin loaded with thorin. *Talanta,* **1999**, *49*(4), 929-935.
[http://dx.doi.org/10.1016/S0039-9140(99)00089-2] [PMID: 18967670]

[212]   Misra, N.L.; Singh Mudher, K.D.; Adya, V.C.; Rajeswari, B.; Venugopal, V. Determination of trace elements in uranium oxide by Total Reflection X-ray fluorescence spectrometry. *Spectrochim. Acta B At. Spectrosc.,* **2005**, *60*, 834-840.
[http://dx.doi.org/10.1016/j.sab.2005.05.023]

[213]   Watanabe, H.; Berman, S.; Russell, D.S. Determination of trace metals in water using x-ray fluorescence spectrometry. *Talanta,* **1972**, *19*(11), 1363-1375.
[http://dx.doi.org/10.1016/0039-9140(72)80133-4] [PMID: 18961191]

CHAPTER 2

Lipidomics Techniques and their Application for Food Nutrition and Health

Shudan Huang[1], Jiamin Xu[1], Jiahui Chen[1], Ting Zhang[1], Jing Su[2], Xichang Wang[1] and Jian Zhong[1,2,*]

[1] *National R&D Branch Center for Freshwater Aquatic Products Processing Technology (Shanghai), Integrated Scientific Research Base on Comprehensive Utilization Technology for By-Products of Aquatic Product Processing, Ministry of Agriculture and Rural Affairs of the People's Republic of China, Shanghai Engineering Research Center of Aquatic-Product Processing & Preservation, College of Food Science & Technology, Shanghai Ocean University, Shanghai 201306, China*

[2] *Xinhua Hospital, Shanghai Institute for Pediatric Research, Shanghai Key Laboratory of Pediatric Gastroenterology and Nutrition, Shanghai Jiao Tong University, School of Medicine, Shanghai 200092, China*

Abstract: Due to the chemical complexity and wide concentration range of lipids in biological samples, it is necessary to apply different analytical strategies to identify and quantify lipid species and amounts. In this book chapter, we mainly introduced the techniques, workflow, and applications of lipidomics in food nutrition and health. First, we mainly introduced the common lipidomics techniques, such as direct infusion mass spectrometry-based techniques, chromatographic separation mass spectrometry-based techniques, mass spectrometry imaging, and nuclear magnetic resonance. Second, we described the common lipidomics workflow, including sample preparation, MS data acquisition, and data processing. Third, we mainly discussed the application of lipidomics in food nutrition and health. Finally, we briefly summarized and discussed the future perspectives of lipidomics. All these discussions suggested that lipidomics could ensure food quality, examine dietary lipid nutrition, and prevent and detect diseases.

Keywords: Chromatography, Cancer, Data processing, Lipidomics, Mass spectrometry imaging, Metabolic syndrome, MS data acquisition, Nuclear magnetic resonance, Nutrition, Neurological disorders, Sample preparation, Shotgun lipidomics.

----
[*] **Corresponding author Jian Zhong:** National R&D Branch Center for Freshwater Aquatic Products Processing Technology (Shanghai), Integrated Scientific Research Base on Comprehensive Utilization Technology for By-Products of Aquatic Product Processing, Ministry of Agriculture and Rural Affairs of the People's Republic of China, Shanghai Engineering Research Center of Aquatic-Product Processing & Preservation, College of Food Science & Technology, Shanghai Ocean University, Shanghai 201306, China; E-mail: jzhong@shou.edu.cn

INTRODUCTION

In recent years, with the development of the economy, the living standards of people have been significantly improved, and people pay much attention to food nutrition and health. Various nutrients play different roles in the body for people. However, food safety and nutrition problems have frequently aroused the panic of people, which might have a bad impact on the lives and health of people. Numerous news has reported the safety and quality of food, such as melamine adulterated milk and [1] gutter oils [2], which improved the concern of the public. Besides, with the globalization of the food market, people can easily get more foods from different countries [3]. Governments and regulatory agencies in the world have issued different legislation and regulations to improve food safety and quality. Although with the persistence of legislation and regulations, there are still amounts of unscrupulous producers and traders who want to produce or sell adulterated foods to earn a big profit, ignoring the health of the consumers [4]. Therefore, we need novel techniques to help to dissolve these problems.

There are lots of nutritional substances in foods, like protein, sugars, vitamins, lipids, inorganic salt, and water [5]. As shown in Fig. (1), there are 9 representative lipids and their reaction products in the food system. Among the nutritional substances, lipids have several key functions, which can be found almost in every cell membrane. The lipid is hardly soluble in water, and very easy to be dissolved inorganic, and has the characteristics of structural diversity and variety [6]. First, it can protect the body from injury and act as a cushion to the body. Second, according to the need, the lipid can insulate the body and keep it warm. Third, it can make the skin and hair lubricated. Moreover, it can help transport the essential fat-soluble vitamins A, D, E, and K and essential fatty acids. Last but not the least, lipid provides energy for people. Each gram of lipids provides 9 calories, which is more than twice that of carbohydrates or protein [7, 8]. In terms of the disease, the lipid has been found to have relations with Alzheimer's disease [9], diabetes [10], tumors [11], and other diseases.

In the 1960s and 70s, lipid was one of the most intensely studied areas of biology. Most of the bioinformatics resources of Kyoto Encyclopedia of Genes and Genomes (KEGG) have relied on the messages, which obtained from that time [13]. In 2003, lipidomics was firstly proposed as one of the main branches of metabolism [14]. It is generally believed that lipidomics can analyze the characteristics of all lipid molecules in organisms and their roles in protein expression and gene regulation. Therefore, the research of lipidomics belongs to the category of life sciences and is closely related to human health, so its importance in the research of diseases has also attracted widespread attention [15]. At present, the contents of lipidomics research mainly include three major aspects: lipid and its metabolite analysis and identification, lipid function and metabolic regulation (including key genes/proteins/enzymes), and lipid metabolism pathways and networks [16, 17].

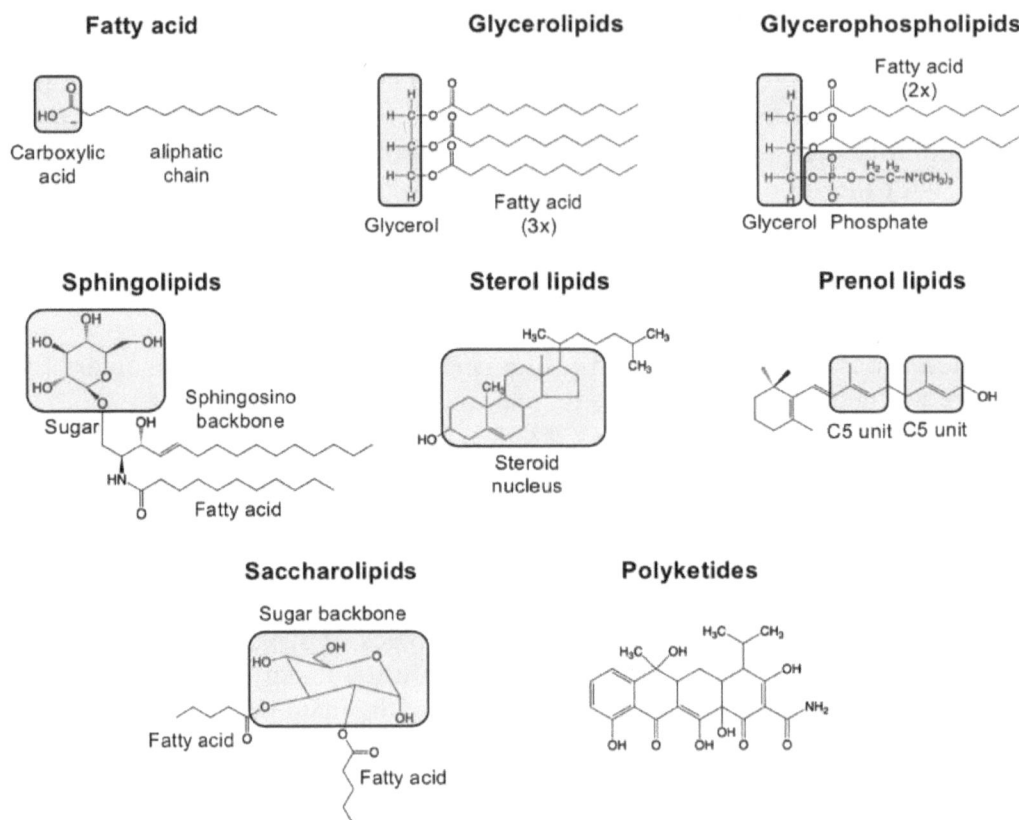

Fig. (1). Categorization of lipids. Reprinted from Elsevier Publisher (2020) [12].

There are several aims of lipidomics, such as to study the lipids in biological fluids, tissues, and cells through various methods, and to explore the body lipids changes in quality metabolism under different diseases or drug interference states, to study the possible mechanism of disease and the mechanism of action of drugs from the perspective of lipid metabolism network, and to search for key lipid biomarkers that can characterize a disease or drug intervention [12, 18, 19].

In this book chapter, we mainly introduce the techniques, workflow, and applications of lipidomics in the fields of food nutrition and health (Fig. **2**). The development of lipidomics techniques can attribute to the applications, such as identifying food adulteration more correctly and detecting diseases quickly.

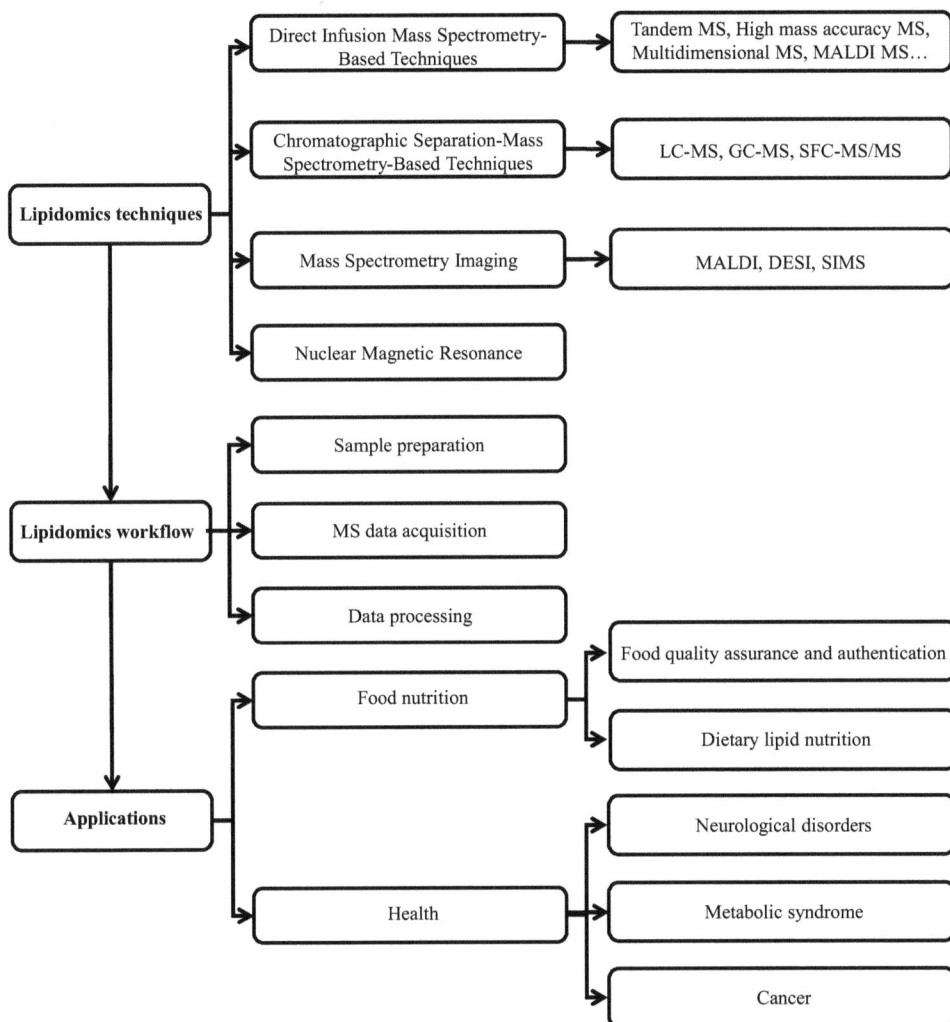

Fig. (2). Workflow of this book chapter.

LIPIDOMICS TECHNIQUES

Because of the chemical complexity and wide concentration range of lipids, which are unveiled in both food and biological samples, it is hard to identify and quantify all lipids simultaneously with a single analytical strategy. As a consequence, according to the different purposes of analysis, lipidomics strategies can be divided into two types: untargeted and targeted approaches. Both of them have their distinctive features, inherent advantages, and limitations [20]. When defined lipids or lipid classes of interest need to be characterized, targeted MS-based approaches are chosen. These approaches have several advantages, like high detection sensitivity, good selectivity, and quantitative accuracy with good linearity and reproducibility [21]. According to the research by Shen *et al.* [22], targeted MS-based lipidomics can serve as the gold standard for lipid quantitation and is suitable for analyses of lipids at low concentrations or with particular structures. Compared with the targeted MS-based approaches, untargeted MS-based lipidomics is a new strategy in lipid analysis. Surprisingly, it can be employed for most lipid classes and can screen lipid markers for food authentication global wide. Due to the high mass resolution (>10, 000) and mass accuracy (<2-5 ppm), high-resolution MS (HRMS), like orbitrap, Fourier transform ion cyclotron resonance (FTICR), time-of-flight (TOF), is often applied [23, 24]. The lipidomics techniques were summarized, as shown in Table **1**.

Table 1. Experimental approaches used in lipidomics.

Technique	Types	Advantages	Disadvantages	References
Direct infusion mass spectrometry-based techniques	Tandem MS	Accurate measurement of the masses of individual molecular ions and fragment ions Broad, efficient, and sensitive measurement Elimination of the possible false-positive identification	Only suitable for the class-specific head group fragments Non-specific MS/MS scanning might lead to the introduction of some artifacts Might be difficult to detect lipid species	[36]
	High mass accuracy MS	Untargeted fashion to analyze any lipid species Applied to many biological studies, lipid species	Different ionization responses of different species among a non-polar lipid, a linear dynamic range of quantification largely depends on the experimental conditions Not suitable for poorly ionized lipids	[36]
	Multidimensional MS	Minimization of the ion suppression effects Use of the peak contours in multi-dimensional space to extend the linear dynamic range	Low throughput and laborious Unable to distinguish isomeric species Not suitable for an unknown or uncharacterized lipid class	[36]

.

(Table 1) cont.....

Technique	Types	Advantages	Disadvantages	References
Direct infusion mass spectrometry-based techniques	MALDI MS	Tremendous sensitivity, high selectivity, and short time High tolerance to salts and other impurities in samples	Multiple adducts complicating spectra Severe matrix background at the low mass range Heterogeneous sample spot and post source decay	[36]
Direct infusion mass spectrometry-based techniques	Ion mobility MS	Separate the many isobaric and isomeric lipids Reduce the complexity of lipid mixtures to simplify the analysis of complex extracts Separate the endogenous matrix interferences to increase the selectivity	Loss of sensitivity, difficult for quantification	[36]
Chromatographic separation mass spectrometry-based techniques	LC-MS	Good reproducibility, high resolving power Tolerant with impurities (like salts) No need for any derivatization reaction of the analytes	Low throughput Prior knowledge required for targeted analysis Retention time drift Effects of mobile phase (*e.g.*, composition, salts) Complex data processing	[29]
	GC-MS	High resolution and sensitivity High temperature (HT)-GC can be applied to analyze some compounds of high boiling	Elimination of much structural information Reduction of throughput	[29]
	SFC-MS/MS	The screen of tremendous samples Maximum efficiency for the target analysis	-	[49]
Mass spectrometry imaging	MALDI	High quality and sensitivity of the appropriate MALDI matrix Suitable for cells, tissues, and even entire bodies	Ion peaks of PC species prominent influence the correct analysis of other positively-charged lipid classes	[42]
	DESI	Provide rapid diagnosis and tumor margin assessment,	Limited sensitivity	[42]

(Table 1) cont.....

Technique	Types	Advantages	Disadvantages	References
	SIMS	No need the prior sample treatment and the closest possible to physiological conditions Give more information about the intracellular correlation	Difficult to detect intact lipids due to severe source fragmentation	[48]
Nuclear magnetic resonance	-	Suitable for pure compounds or complex mixtures Quantify the lipids	Limited sensitivity High cost in data analysis Cannot identify metabolites in complex mixtures	[50]

Direct Infusion Mass Spectrometry-based Techniques

According to the station of the lipid solution, whether it is transported to the ion source chamber without prior separation, MS-based lipidomics approaches can be divided into two categories, direct infusion-based lipidomics (also named shotgun lipidomics) and chromatograph-MS-based lipidomics. In 2005, Han *et al.* firstly put forward the concept of shotgun lipidomics [25], which has the advantages of quantitative hundreds of lipid species in a relatively short time correctly. However, due to the absence of pre-chromatographic separation, the lipid extract is complex.

In order to detect molecular species selectively and do not interfere with other coexisting classes, MS/MS can monitor head-group related fragments, which are only existed in lipids. There are several tandem mass spectrometric techniques used in lipidomics: Precursor Ion Scan (PIS), Neutral Loss Scan (NLS), and Selected/Multiple Reaction Monitoring (SRM/MRM). The first mass analyzer of PIS can scan all the precursor ions; the second mass analyzer monitors can only scan the selected fragment ion. The precursors produced during fragmentation are monitored because the selected fragment ion corresponds to a common fragment [26]. Compared with the PIS, the first mass analyzer of NLS scans all the precursor ions, while the second mass analyzer scans the fragment ions that are offset from the first mass analyzer [27]. Due to the offset leading to neutral loss, all of the precursors that experience the loss of the specified neutral fragment are supposed to be monitored. SRM/MRM is different from the PIS and NLS; it is a non-scanning tandem mass spectrometric technique performed on triple quadrupole-like instruments [28]. This technique uses two mass analyzers, acting as a static mass filter, to monitor a particular fragment ion of a selected precursor ion [29].

The principle of shotgun lipidomics maximizes utilization of the unique chemical and physical properties of lipid classes, sub-classes, and individual molecular species to directly identify and quantify the cellular lipids from organic extracts of biological samples [30]. An important feature of shotgun lipidomics is that this principle is only effective when the lipid solution is under a constant concentration. The factors that influence the identification and quantification of individual lipid species can be eliminated under a constant concentration. Getting a full mass spectrum to display the molecular ions of all the species of a lipid class of interest is another major feature of shotgun lipidomics. Thanks to this feature, molecular species can be easily visualized, and scientists are allowed to identify and quantify the PIS and NLS without time constraints [12].

Due to the unique characteristics, there are five different approaches of shotgun lipidomics, including tandem MS, high mass accuracy MS, multi-dimensional MS, MALDI MS, and ion mobility MS.

The key to success in the tandem MS is the specificity of the characteristic fragment. This approach has the advantages of efficiency, less expensive instrumental requirements, ease of management, simplicity, and high sensitivity. Tandem MS can provide global determination for any targeted class at the level of instrumentation sensitivity [31 - 33]. However, there are some concerns. First, because this method only targets the class-specific head group fragments thus, there is no fatty acyl substitution of lipid species. Secondly, non-specificity might lead to the introduction of some artifacts, and the results might not be accurate. Moreover, it is not easy to detect the lipid species due to the differential fragmentation thermodynamics and kinetics manifest in individual lipid species within each lipid class [34].

Commercially available hybrid type mass spectrometers can accurately measure the masses of individual molecular ions and fragment ions due to advanced instrumentation and high mass accuracy MS (*e.g.*, 0.1 amu or higher). The molecules, as well as fragment ions, can eliminate the possible false-positive identification [23, 35]. This method of shotgun lipidomics has the advantages of providing broad, efficient, and sensitive measurement of lipid species in a high-throughput fashion, analyzing any lipid species present in cellular lipids if the dynamic range of the instrument is allowed. Besides, this method can be applied to all those species in an untargeted fashion [36].

While using the differential, stability, sub-classes, hydrophobicity, and re-activity of distinct lipid classes when preparing the sample, multi-dimensional MS-based shotgun lipidomics (MDMS-SL) can analyze the lipid under very low abundance levels of molecular species [37]. There are several advantages of multi-dimensional MS. First, using a mass spectrometer as a separation tool can minimize the ion suppression effects. Second, employing multi-building blocks of individual molecular species can identify the structure and isomers [38]. In this way, the presence of any artificial species can be eliminated. Moreover, the multidimensional MS can be applied to exploit the distinctive chemical characteristics of many lipid classes. However, there are also some drawbacks to this method. Due to the involvement of different procedures (*e.g.*, derivatization) in multiplexed sample preparation, the throughput is relatively low and laborious. Second, the fragmentation patterns are the same as other shotgun lipidomics approaches. As a consequence, multi-dimensional MS cannot distinguish isomeric species. Also, this approach is not an ideal way to identify and quantify the unknown or uncharacterized lipid class since the identification of the building blocks of a lipid class has to be pre-determined [36].

The approach of MALDI MS has the features of tremendous sensitivity and high selectivity and short analysis time. Besides, it has a high tolerance to salts and other impurities in samples. Easy sample preparation, high throughput, and speed re-analysis of samples are the advantages of the MALDI MS. Multiple adducts complicating spectra, severe matrix background at low mass range, heterogeneous sample spot, and post source decay are the limitations of this method [39]. In the case of ion mobility MS, the ionized molecules are separated by size, shape, charge, and mass according to their ion mobility in low or high electric fields. There are three advantages of ion mobility MS. First, it can separate the many isobaric and isomeric lipids, which are difficult to be assessed by other shotgun lipidomics. Second, it can reduce the complexity of lipid mixtures to simplify the analysis of complex extracts. Finally, it can separate the endogenous matrix interferences to increase selectivity [40, 41].

Chromatographic Separation-mass Spectrometry-based Techniques

With a unique high separation capability, chromatography is a basic tool to analyze complex samples in a comprehensive way. There are various chromatographic methods, like high-performance liquid chromatography (HPLC), thin-layer chromatography (TLC), capillary electrophoresis (CE), gas chromatography (GC), and supercritical fluid chromatography (SFC). These methods can meet the demands of tremendous lipidomics analysis. Especially when hyphenated to MS, chromatography methods can provide more information on the lipidome [42]. In comparison to other approaches, the chromatograph-M--based lipidomics analysis allows the clear and specific lipid species to enter the mass spectrometer at different times during the separation process before the MS analysis. Hence, it can detect lipids under an extremely low abundance [42]. However, it needs more analysis time than the shotgun lipidomics.

The most frequently-used separation technique in lipids analysis is liquid chromatography (LC) that has the advantages of good reproducibility and high resolving power. Besides, LC has tolerance with impurities (like salts) and does not need any derivatization reaction of the analysis [15]. As for HPLC, because the feature of HPLC systems is relatively isolated from the environment, the contact between samples and air is reduced, which avoids the sulfoxidation and degradation of lipids. LC configurations can be mainly divided into two categories: reverse phase (RP)-LC and normal phase (NP)-LC, which have different purposes for lipidomics analysis. Separating different classes of lipids based on the polar head groups is using NPLC. Separating different molecular species in one class based on the different fatty-acyl chains is using RPLC. This method has the advantages of reduction of ion suppression, highly sensitive targeted analysis, separation of isomers and isobars [43]. However, the effects of the mobile phase (*e.g.*, composition, salts, or additives), complex data processing, and low throughput are the limitations of this approach. As an important complementary method to LC, gas chromatography can separate some lipids like FAs and STs. According to Chen *et al.*, GC and GC-MS are promising technologies for lipidomics by analyzing fatty acid chains [44]. The advantages of GC are high resolution and sensitivity. High temperature (HT)-GC can be applied to analyze some compounds of high boiling points.

There are also some drawbacks of GC, like the elimination of much structural information and reduction of throughput. SFC, a high-resolution technique, can be employed to separate the lipids. SFC-MS/MS methods can be applied to the screening of tremendous samples. Recently, SFC-MS has shown the maximum efficiency for the target analysis [45]. Because the majority of lipids are hydrophobic and have low UV absorbance, CE with UV detection is rarely used in lipid analysis. Therefore, Gao *et al.* reported a CE-UV method, which has the characteristic of acceptable reproducibility and relatively low sensitivity [46].

Mass Spectrometry Imaging

After the extraction and purification steps, the information about the spatial distributions of lipids throughout heterogeneous tissue or cell biological samples will be lost. Therefore, mass spectrometry imaging (MSI) can be used as an imaging technique to visualize the dynamic spatial distributions of different lipid classes within biological tissues and identify the changes of localized lipid composition and interaction with adjacent microenvironments [47].

Matrix-assisted laser desorption ionization (MALDI), as same as electrospray ionization (ESI), is one of the earliest ionization techniques applied in lipidomics analysis. The high quality and sensitivity of the appropriate MALDI matrix is the key feature of MALDI. It also has the characteristic of application in the imaging lipid profiles of cells, tissues, and even entire bodies. This technique can visualize the lipid distribution and relative concentrations, but the ion peaks of PC species prominent influence the correct analysis of other positively-charged lipid classes [42]. Desorption electrospray ionization (DESI) is important to the lipidomics investigation. This technique is applied in the diagnose of some cancers. Providing rapid diagnosis and tumor margin assessment, DESI makes the near-real-time tumor surgery guide possible. But, the limited sensitivity is the drawback of this technique [42]. The secondary ion MS (SIMS) was first reported in 1971 but used in the lipidomics field much later than ESI and MALDI. This method can be used without the need for prior sample treatment. Hence, SIMS can provide the closest possible physiological conditions. Moreover, it has the advantage of high spatial resolution (submicrometer-grade). Single-cell detection can provide more details of some biological procedures, which are crucial to the cellular processes. Moreover, it can give more information about intracellular correlation [48].

Nuclear Magnetic Resonance

Due to the no-invasion of the sample and short time-consuming, NMR spectroscopy (1H, 2H, and 31P) arouse the interest of the scientist in metabolomics [51]. Regardless of any previous treatment, nuclear magnetic resonance is helpful to analyze the metabolites in a single sample. It is a widely-used method and can be employed in pure compounds or complex mixtures. However, compared with MS, the sensitivity of NMR is low; the spectra of complex mixtures are hard to explain for the superimposition of the metabolite signals. The disadvantage of high cost in data analysis stops employment. The signal intensity of each resonance in the NMR spectrum is directly proportional to the number of spins associated with the particular resonance. This characteristic helps to quantify the lipids [52]. Although the NMR has a good capability of quantification, it cannot always identify metabolites in complex mixtures. As a consequence, NMR is used less in lipidomics [50].

LIPIDOMICS WORKFLOW

In order to use the lipidomics approaches to understand the functions of lipids, it is necessary to optimize the workflow, such as sample collection and preparation, derivatization, LC separation, MS analysis, quality control, data processing, and data interpretation. A typical workflow of lipidomic analysis of biological samples can be divided into sample preparation, MS data acquisition, and data processing. It is supposed to prepare the sample before the instrumental analysis for getting accurate results. The aim of sample preparation is to separate lipids from the contaminating molecules in a biological or food matrix. As a consequence, a lipid-enriched sample is suitable for further analysis. Manual lipid extraction is prone to error and is labor-intensive, resulting in the impediment of large-scale studies with massive samples. Therefore, large-scale studies need to automate the sample preparation and extraction process. Improving the cost-effectiveness of the study is the advantage of automation. It is supposed to maintain the endogenous lipids intact and pay attention to the reagents, sample amounts, protocols, lipid standards, hardware, and solvent for all components that will impact the accuracy of the lipidomics dataset.

There are several common extraction methods. Modified Bligh and Dyer method is a well-established standard method and has been practiced broadly: Chloroform/methanol/H_2O (1:1:0.9, v/v/v) for extraction of a small amount of biological sample. After phase separation, total lipids are present in the chloroform phase. Using hazardous chloroform and collecting chloroform extract from the bottom layer are the drawbacks, which will lead to the impurity of the water-soluble and automation difficulty. The modified Folch method has similar advantages and disadvantages to the modified Bligh and Dyer method. Firstly, this method employs chloroform/methanol (2:1, v/v) to extract biological tissue, then add water or 0.9% NaCl (0.2 volume) to wash the solvent extract. The methyl tert-butyl ether (MTBE) method resolves some of the difficulties in chloroform-involved methods because the MTBE is present in the top layer after phase separation. This method is capable of high throughput as well as automation. Containing a significant number of aqueous components that may carry over water-soluble contaminants is the limitation of this method. The butanol/methanol (BUME) method can reduce the water-soluble contaminants carried over in the organic phase. The method is to add a volume of BUME (3:1, v/v) to a small volume of an aqueous phase, add an equal volume of heptane/ethyl acetate (3:1, v/v), then add 1% acetic acid (equal volume to BUME) to induce phase separation. However, the butanol component in the organic phase is hard to evaporate [29].

After extraction, there are some optional steps before the MS analysis. The direct infusion-based shotgun approach relies on reducing the complexity of the extract in order to avoid the need for separation or enrichment prior to MS analysis. It can employ physical approaches, like solid-phase extraction or liquid/liquid partitioning to separate polar or nonpolar lipids [50]. Besides, chemical approaches, like base hydrolysis, can enrich the low abundance sphingolipid from complex lipid extracts that contain high abundance phospholipids and/or glycerolipids [53].

Direct infusion mass spectrometry-based techniques, chromatographic separation-mass spectrometry-based techniques, and mass spectrometry imaging are usually applied in the lipidomics analysis. There are some MS ionization techniques, such as matrix-assisted laser desorption/ionization (MALDI), atmospheric pressure chemical ionization (APCI), atmospheric pressure photoionization (APPI), secondary ion mass spectrometry (SIMS), and desorption ESI (DESI). After the lipid ionization, it is optional to mobilize the ion before the MS analysis. Then, full MS or MS/MS analysis, or both, can be applied, which depends on whether a targeted or global analysis is desired. Following MS analysis, the data can be shown as the ion chromatogram, extracted ion chromatogram, MS spectra, MS/MS spectra, and images.

Recently, MS analysis is usually used in lipidomics analysis. The raw binary collected data from MS instruments should be obtained first. The lipidomic dataset should be in the form of a matrix, with molecular lipids representing a significant portion of the total lipidome. Besides, in each sample, the lipidomic dataset should be absolutely quantified. However, according to different analytical approaches and types, it is necessary to trade-off. In the process of LC-MS data, peak detection and integration, alignment, identification, and normalization are needed [14]. Wu *et al.* used the lipidSearch software to study the effect of dry salting on the long-chain free fatty acid profile of tilapia muscles. The results showed that lipid search software could be used for the analysis of the effect of dry salting on the free fatty acids [54]. Shi *et al.* used the lipidSearch software to study the effects of thermal processing methods on food lipids (Fig. **3**). After different thermal processing methods, the lipidomics profile shows differences (Fig. **4**). This work could provide useful information for aquatic product processing and lipidomics [55].

After processing the data, the lipids can be studied by various statistical methods that are dependent on the experiment design and the questions asked. Univariate and multivariate methods are employed in the lipidomics data analysis. Due to the similar function and structure between the measured lipids, the multivariate approach serves as a natural method to study and represent the lipidomics data. The molecular detail acquired in the lipidomic analysis is suitable for most lipid classes and beyond the detail level gotten from the biochemical pathway. As a result, it is important to use a new tool to promote the visualization and reconstruction of lipid at the molecular level [56]. Some commercial software such as metal analyst software can be applied for statistical analysis [57, 58] in which the data were analyzed by partial least squares-discriminate analysis (PLS-DA), variable importance in projection (VIP) scores, and heat map visualization. Shi *et al.* showed the statistical analysis of raw and thermal processed tilapia fillets using metal analyst software (Fig. **5**), which proved the significant differences among these samples.

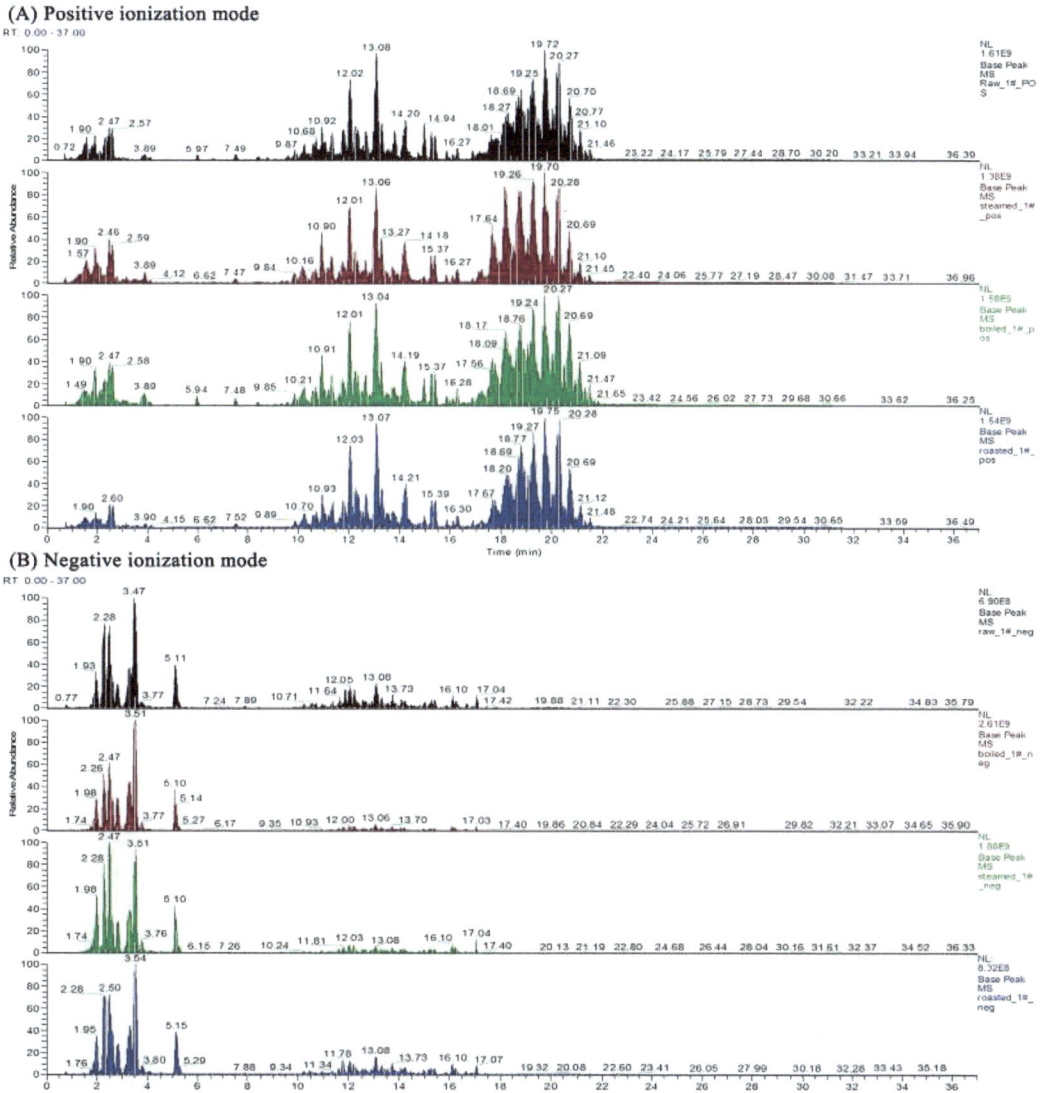

Fig. (3). Typical UHPLC-Extracted MS base peak intensity chromatographs of tilapia fillets. The spectra from top to bottom are as follows: raw, steamed, boiled, and roasted tilapia fillets. Reprinted with permission from reference [55] (Elsevier Publisher 2019).

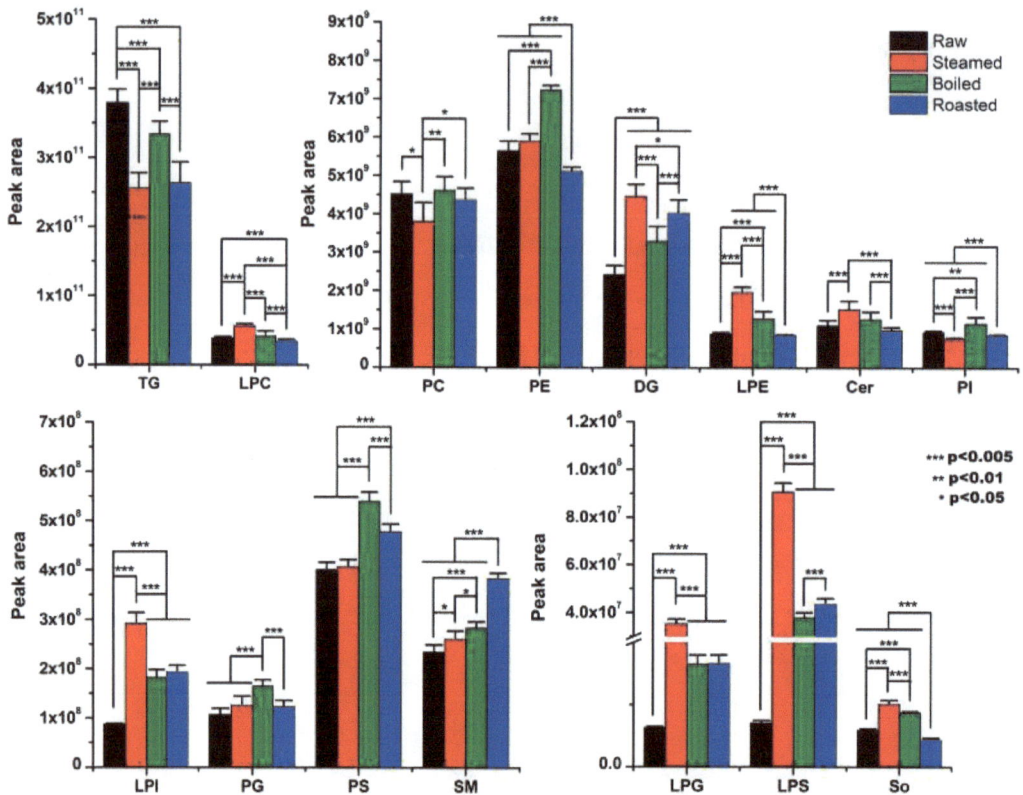

Fig. (4). The analyzed peak areas in raw and processed tilapia fillets (n=6) by using commercial LipidSearch lipidomics software. Reprinted with permission from reference [55] (Elsevier Publisher 2019).

Finally, the lipidome is modeled in the physiological context. In the spatial and temporal context, the lipidomic data as a part of the biological system should be studied. Otherwise, there are still lots of challenges left. The computational complexity needs to be reduced. Even though, after the simplifications, it takes tremendous time to simulate the data. Therefore, it is impossible to achieve the aim of routine lipidomic data analysis. In the case of lipoprotein modeling, it is crucial to think about the availability of lipidomic data from different lipoprotein fractions. No matter whatever limitations exist in lipidomics, the applications of computational lipidomics have brought crucial biological insights and new hypotheses [14].

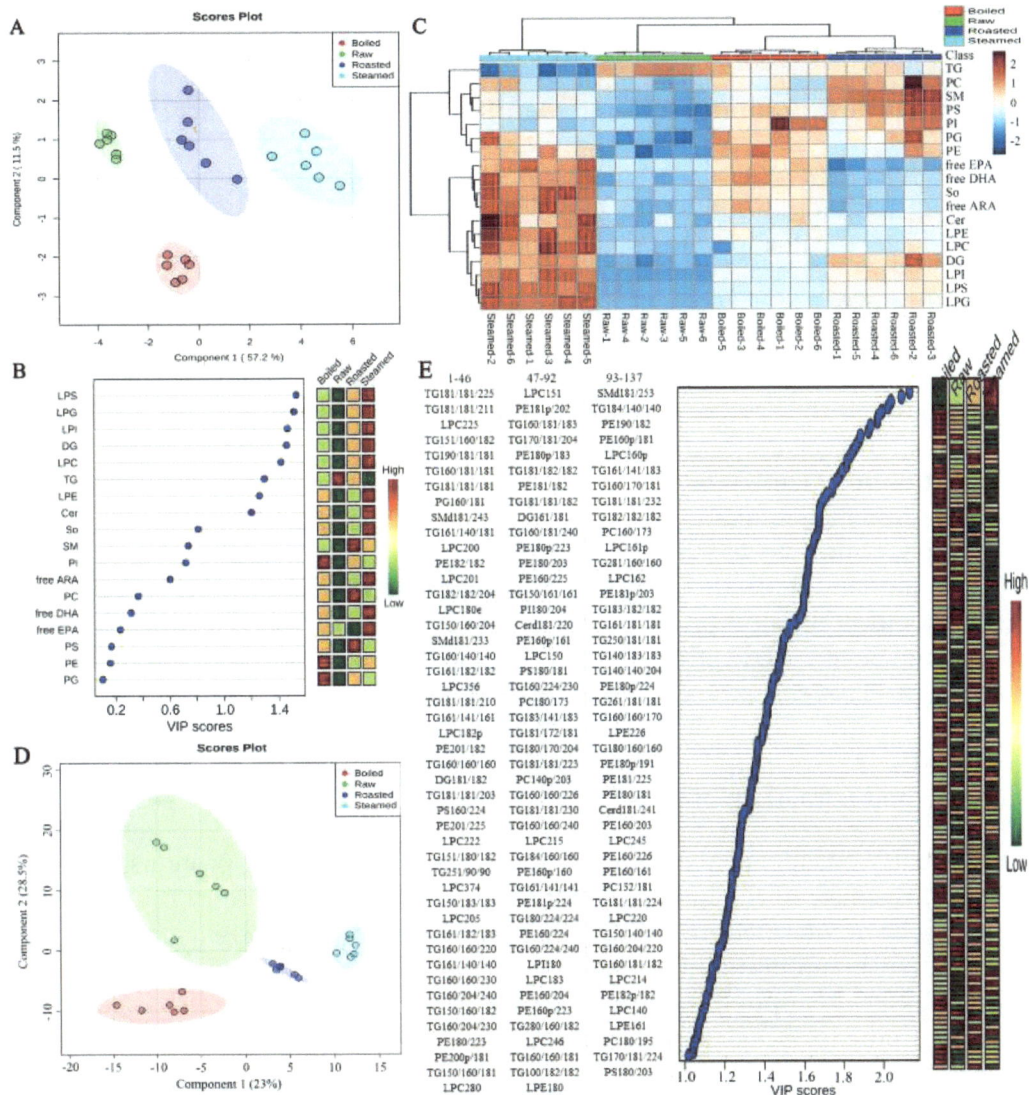

Fig. (5). Statistical analyses of lipids in raw and thermal processed tilapia fillets by Metaboanalyst 4.0 software. **(A):** PLS-DA score plot of lipid species. **(B):** VIP scores of lipid species in PLS-DA. **(C)** Heat map of lipid species. **(D)** PLS-DA score plot of individual lipids. **(E)** VIP scores of individual lipids in PLS-DA. The lipid order from top to down is shown on the left. Reprinted with permission from reference [55] (Elsevier Publisher 2019).

APPLICATIONS OF LIPIDOMICS

Applications of Lipidomics in Food Nutrition

Lipidomics, as one of the omics, can provide a special perspective for solving the problem in foodomics. This part demonstrates the applications of lipidomics in food nutrition: food quality assurance and authentication and dietary lipid nutrition.

Applications of Lipidomics in Food Quality Assurance and Authentication

The lipidomics can identify the different species, geographical origins, and food processing methods. Foods with high lipid amounts, such as edible oils, milk, dairy products, meat, and kinds of seafood, are easy to be adulterated.

Edible oils consist of TAGs and a trace amount of diacylglycerols (DAGs), free fatty acids (FFAs), PLs, and other minor components. They are crucial dietary sources of lipids for humans [59]. TAGs are the main discriminating markers for authentication, also serve as a key indicator in edible oil quality assessment. Hou *et al.* unveiled that LF-NMR combined with SVM can be used for edible oil identification for identifying the edible oil species based on the low-field nuclear magnetic resonance relaxation features. In addition, the accuracy of this method is quietly high, almost up to 99.04%. The advantage of this method is fast and needs the minimum sample without producing chemical waste, and the entire feature extraction and prediction process only costs 144 s [60]. Ng *et al.* established the first comprehensive MALDI MS spectral database and divided the edible oils into eight groups. This research used the partial least square-discriminant analysis to analyze 435 edible oil products with an overall accuracy of 97.2%. Besides, the MALDI MS spectra could be used for the classification of cooked edible oils, gutter oils, and normal edible oils [61].

Milk and dairy products have many nutrients, like fats, lactose, proteins, vitamins, and so on. Because of the high nutritional values of milk and dairy products, they are easy to be adulterated. In the adulteration, vegetable oils (*e.g.*, soybean, palm, and peanut oil), animal fats (lard and tallow) are added to improve the content of lipid. Hrbek *et al.* applied the DART-HRMS to study the authentication of milk and dairy products. Besides, the research also found that it is of high efficiency to detect the exogenous oils of plant origin. Surprisingly, the level of sunflower, rapeseed, and/or soybean oil was as low as 1% (w/w) could be detected. Moreover, it only needed a minimal sample. Additionally, the oxidation of plant TAGs could be a marker for adulteration with poor-quality vegetable oils.[62] Rebechi *et al.* detected the adulterations of milk fat with animal fats by applying MLR data processing and GC analysis of FA profiles of MF jointly. This research proved that multiple linear regression is a valid statistical tool for the evaluation of adulterations of milk fat. In addition, this method can detect the adulterate sample with 5% lard and 10% tallo [63]. The detection of tallow is the most difficult thing in the adulteration study.

In 2013, the horse DNA was found in frozen beef burgers in Europe, which aroused the concern of people. It is unveiled that minced and homogenized meat is easy to be replaced by other cheaper meat. Due to the high proportion of lipid in these foods, lipidomics can be applied for authentication. One of the markers for discrimination of wild or conventionally farmed fish is FA composition. Trocino *et al.* compared the organic and conventional sea bass and unveiled that organic sea bass had high n-3 to n-6 PUFA ratio than conventional sea bass. Production systems could not influence the composition of the organic and conventional sea bass. However, the diets fed would affect the lipid composition and the dietary value of muscles, and the commercial size of the sea bass would impact the biometric traits without affecting the dietary value of flesh. Hence, it was a potential method to identify the sea bass [64]. Fiorino *et al.* developed a DART-HRMS method to discriminate wild-type from farmed salmon. This method used the principal component analysis (PCA) to integrate and then apply to scout for spectral features that are helpful to discriminate wild-type from farmed salmon of Salmo salar species. The research found that three saturated (14:0, 16:0, and 18:0) FA, along with unsaturated ones having 20 or 22 carbon atoms, were the differences between wild-type and farmed salmon. Besides, FA with compositions 18:1, 18:2, 18:3 in farmed salmon was higher than wild-type salmon [65].

Applications of Lipidomics in Dietary Lipid Nutrition

For studying the interactions between genes, diet, nutrients, and metabolism, lipidomics has been applied in food and nutrition research. Omega-3 fatty acids can be used for the prevention and/or therapeutics. Typically, DHA and EPA can reduce TG and inflammation [66, 67].

According to the study by Lamaziere *et al.*, long-chain omega 3 fatty acids could be electively incorporated into hepatic phospholipids, inhibit de novo lipogenesis, and change the hepatic fatty acid profile by reducing the activity of desaturase in the non-steatotic liver. Moreover, the data in this research is helpful for the hepatic fat and ameliorates nonalcoholic fatty liver disease (NAFLD) prognosis [68]. Recently, more and more evidence demonstrated that lipids correlated with the regulation of intestinal immunity and biophylaxis and immune disorders. According to Kunisawa et al., sphingolipids can act directly on IgA antibody-producing cells and increase the generation of IgA antibodies, and 17,18-EpETE could be the active molecule in the EPA-derived lipid mediator that inhibited the intestinal allergy [69]. Heilbronn *et al.* studied the influence of short-term overfeeding. The report indicated that, after short-term overfeeding, the high-density lipoproteins would increase, but the low-density lipoproteins, triglycerides, and non-esterified fatty acids would keep unchanged [70]. Rey *et al.* analyzed the influence of n-3 long-chain polyunsaturated fatty acid (LC-PUFA) supplementation on oxylipin profile and neuroinflammation in the brain. They found that hippocampal oxylipins profile could be adjusted by n-3 (LC-PUFA), which also could reduce the production of hippocampal pro-inflammatory cytokines. The results suggested that dietary habits might impact the production of brain cytokine [71].

Applications of Lipidomics in Health

Applications of Lipidomics in Neurological Disorders

The highest amounts of lipids are in the brain, and the lipids have associations with neurological disorders. In the aging population, dementia is usually caused by Alzheimer's disease (AD). Preclinical and clinical evidence demonstrated that docosahexaenoic acid (DHA), an omega-3 fatty acid derived from the diet or synthesized in the liver, could reduce the risk of developing Alzheimer's disease. Astarita and Piomelli unveiled that, due to the deficit in the peroxisomal D-bifunctional protein, the liver of AD patients could only convert a little shorter-chain omega-3 fatty acids into DHA. As a result, this might lead to decreased DHA levels in the brain and cognitive impairment [72]. The research conducted by Proitsi *et al.* employed lipidomics to identify a battery of plasma metabolite molecules for predicting AD patients from controls. The researchers unveiled the new molecules, which could diagnose AD patients from controls and MCI, by untargeted lipidomics method. Moreover, it could improve the recognition of AD and its underlying biology. The Random Forest approach was applied in this study and found the combination of 10 metabolites could predict AD with near 80% accuracy. Moreover, the precursor of ChE (cholesterol) had no direct correlations with AD, but new ChE molecules had correlations with AD, which might be helpful for AD therapy [73]. Wood *et al.* unveiled differences in lipidomics of the mild cognitive impairment (MCI) and old dementia (OD) cohorts. With the aggravation of the disease, monoacylglycerols (MAG), diacylglycerols (DAG), and the very long-chain fatty acid 26:0 would increase. The Ethanolamine plasmalogens in young dementia (YD) and OD cohorts and the phosphatidylethanolamines (PtdEth) in the MCI, YD, and OD cohorts both decreased. However, the ethanolamine plasmalogens decreased in the YD and OD groups, whereas they did not decrease in the MCI group. Therefore, the composition of ethanolamine plasmalogens would not be the symbol of transition from MCI to frank dementia [74].

Huntington's disease (HD) is a neurodegenerative disease caused by a CAG expansion in the HD gene, which encodes the protein Huntingtin. Vodicka *et al.* verify the method of LC-MS/MS to identify changes in lipids and metabolites in HD animal models [75]. Multiple sclerosis is a neurodegenerative disease of the central nervous system (CNS) and one of the most common neurological (non-traumatic) reasons for disability in young adults. The lipid profile might be the symbol of this pathology and lead to dysregulation of lipid homeostasis and lipid metabolism in multiple sclerosis. Specifically, fatty acids 18:2 and 20:4 and total polyunsaturated FA decreased, with compensatory increases in saturated FA with shorter carbon chains. Besides, clinical lipidomics can analyze the lipids for multiple sclerosis patients [76].

Applications of Lipidomics in Metabolic Syndrome

Due to the risk factors, metabolic syndrome will increase the potential of metabolically related diseases, like cardiovascular diseases, diabetes, stroke, and nonalcoholic fatty liver disease. Besides, with the level of obesity increasing, more and more people easily get metabolic syndrome [77]. Cardiovascular diseases (CVD) account for approximately 30% of all deaths in the world [78]. One of the main factors affecting cardiovascular diseases is dysregulated lipid metabolism. Prior to the incident of CVD, atherosclerotic plaques can usually be observed. Zhang *et al.* found PM2.5 led to inflammation along with cholesterol esters, glycerophospholipids, glycerolipids, and sphingolipids metabolism disruption. All of them had relations with atherosclerosis. Besides, this influence would last for 1 month [79]. Yan *et al.* firstly used the fit-for-purpose validations of the quantitation method by UPLC-QTRAP-MS/MS to guarantee the reliability of the results. After integrating typical static metabolomics and time-dependent analysis, the researchers found twelve Lyso-PLs could change with the development of atherosclerosis. Hence, it could serve as a reliable biomarker [80]. Vascular calcification has correlations with conventional cardiovascular risk factors, such as hypertension, diabetes mellitus, body mass index, race/ethnicity, and family history of myocardial infarction. Vorkas *et al.* studied the serum samples from different levels of calcific coronary artery disease (CCAD) based on their calcium score (CS), applying the lipid profiling and complementary modeling approaches. The results suggested that the symbol of disturbing inflammation homeostasis was the level change of 18-carbon and 20:4 FAC lipids [81].

Type-1 diabetes mellitus (T1DM) is an immune disease. Due to the β-cells being destroyed, the patients have difficulty in secreting insulin and controlling the blood glucose. Sobczak *et al.* measured the plasma concentrations of major FFA species in individuals with T1DM and an age/sex-matched control group without diabetes. The results showed that age and sex had no obvious association with the plasma FFA concentrations of T1DM. However, elongase is negatively correlated with age (p=0.0363). Plasma FFAs had obvious relations with cholesterol and

HDL, but not with LDL or diabetes duration. This study made progress in understanding T1DM and its effects on lipid metabolism [81].

Applications of Lipidomics in Cancer

Lipids play an important role in all the basic processes of tumor development. Lipids are necessary for the proliferation of cancer cells; non-esterified fatty acids are key compositions for lipid biosynthesis, and remodeling, bioactive lipids (*e.g.*, lysophospholipids or oxidized lipids) are crucial to the transition of signal, and so on. Therefore, lipidomics can be applied in cancer diagnosis and therapy [29].

Cervical cancer (CC), as one of the most common female genital tract tumors, can be seen commonly in low-income countries. Although HPV vaccines have been used to prevent HPV infection, there are still a tremendous number of people who cannot obtain the immunization programs for the high cost of HPV vaccines. Hence, we must pay attention to the early CC screening. Cheng *et al.* employed the coupling ultra-high-pressure liquid chromatography (UHPLC) with quadrupole time-of-flight tandem mass spectrometry (Q-TOF-MS) to analyze the sample by the non-targeted lipidomics study. Receiver operating characteristic (ROC) curve and binary logistic regression were applied to assess the diagnostic potential of these biomarkers and then to get the best biomarker combination. This study demonstrated that the combination of PC 14:0/18:2, PE 15:1e/22:6, and PE 16:1e/18:2 could act as the promising serum biomarker to discriminate early-stage CC from SIL and healthy subjects [82].

Prostate cancer is the second reason for cancer deaths in American men [83]. Bedia *et al.* employed the untargeted lipidomic approaches to analyze the sample. In the prostate cancer cell line DU145, this research reported that the aldrin, Aroclor, and CPF would induce a pro-metastatic phenotype. Moreover, the lipids might attribute to the malignant phenotype [84]. Osteoarthritis (OA) is one of the most common types of chronic arthritis, and more and more reports indicated that the lipids impacted OA. Rocha *et al.* elucidated the heterogeneity and spatial distribution of lipids in the OA synovial membrane; they also discussed the relation with inflammation. This study reported that compared with the control, the level of phosphatidylcholines, fatty acids, and lysophosphatidic acids were elevated, while lysophosphatidylcholines were low. Hypertrophic, inflamed, or vascularized synovial areas impacted the spatial distribution of particular glycerophospholipids [85]. Sun *et al.* used the ultra-performance liquid chromatography-tandem mass spectrometry (UPLC-MS/MS) to compare the tissues from the human gastric cancer and adjacent normal from clinical patients. The results indicated that, in AGS, SGC-7901, and MGC-803 gastric cancer cell lines, the palmitic acid (PA) downregulated greatly, and the high concentration of

PA was hindering cell proliferation. In addition, sterol regulatory element-binding protein 1 (SREBP-1c) could promote the expression of a series of genes associated with the synthesis of fatty acids. As a consequence, controlling the SREBP-1c could be a potential method to treat gastric cancer [86]. The application examples of lipidomics in health discussed above are summarized in Table **2**.

Table 2. Applications of lipidomics in health.

Application	Type	Technique	Analyzed Molecule	Finding	References
Neurological disorders	Alzheimer's disease	LC/MS	DHA	Due to the deficit in the peroxisomal D-bifunctional protein, the patients had a limitation in converting shorter-chain omega-3 fatty acids into DHA.	[72]
		HPLC	ChE	The cholesterol (the precursor of ChE) had no relation with AD. However, the ChE molecules, which participates in the cholesterol metabolism, had an association with AD.	[73]
		High-resolution mass spectrometric	MAG, DAG, very-lon--chain fatty acid 26:0	Common underlying pathophysiology in the MCI and OD groups. Alterations in plasmalogen synthesis were unlikely to represent an initiating event in the transition from MCI to frank dementia.	[74]
Neurological disorders	Huntington's disease	LC-MS/MS	Glycerophospholipids including alterations in phosphatidic acid, lysophosphatidic acid	In a knock-in HD mouse model, the specific species of lipids changed, and phospholipids were important in neurodegenerative disease.	[75]

(Table 2) cont.....

Application	Type	Technique	Analyzed Molecule	Finding	References
Metabolic syndrome	Cardiovascular diseases	UPLC-QqQ-MS/MS	Glycerophospholipids, glycerolipids, sphingolipids	Atherosclerosis had an association with PM2.5 induced inflammation along with cholesterol esters, glycerophospholipids, glycerolipids, sphingolipids metabolism disruption.	[79]
		UPLC-QTRAP-MS/MS	Lyso-PC/15:0, 18:1/Lyso-PI, 22:5/Lyso-PI, 22:4/Lyso-PI	Twelve lyso-PLs had relations with the progress of atherosclerosis, and hierarchical clustering analysis based on these lyso-PLs could differentiate the three animal groups.	[80]
	Type-1 diabetes	GC-MS	Saturated/unsaturated FFA ratios, n-3/n-6 FFA ratio, de novo lipogenesis index, and elongase index	In T1DM, the concentrations of most major FFA species reduced, the saturated/unsaturated FFA ratios, n-3/n-6 FFA ratio, de novo lipogenesis index, and elongase index increased, SCD1 indices 1 and 2 decreased.	[81]
Cancer	Cervical cancer	UHPLC	Phosphatidylcholine, phosphatidylethanolamine	The combination of PC 14:0/18:2, PE 15:1e/22:6, and PE 16:1e/18:2 could act as a serum biomarker for dividing the early-stage CC from SIL and healthy subjects.	[82]
	Prostate cancer	LC-MS	TAG, LPA, SLs	Lipidic signatures could better understand the lipid metabolic pathways and had relations with EMT and metastasis, which was helpful to the therapeutic targets in cancer research.	[84]

(Table 2) cont.....

Application	Type	Technique	Analyzed Molecule	Finding	References
Cancer	Synovial inflammation	MALDI-MSI	Phosphatidylcholines, fatty acids, lysophosphatidic acids, lysophosphatidylcholines	Levels of phosphatidylcholines, fatty acids, and lysophosphatidic acids were higher, and levels of lysophosphatidylcholines were lower in OA. The spatial distribution of particular glycerophospholipids was related to the hypertrophic, inflamed, or vascularized synovial areas.	[85]
	Gastric cancer	UPLC-MS/MS	Palmitic acid, SREBP-1c	Fatty acid synthesis was different in gastric cancer, and SREBP-1c could act as a promising target for gastric cancer treatment.	[86]

CONCLUSION AND FUTURE PERSPECTIVES

As discussed above, lipidomics was firstly proposed as one of the main branches of metabolomics in 2003 and had relations with food nutrition and health. There are several techniques applied in the analysis of the sample, such as direct infusion mass spectrometry-based techniques, chromatographic separation-mass spectrometry-based techniques, mass spectrometry imaging, and nuclear magnetic resonance. Those approaches can improve the recognition of the correlation of lipid, provide more detailed information, and apply in more fields. To improve the lipidomics approaches, it is necessary to optimize the workflow, including sample collection and preparation, derivatization, LC separation, MS analysis, quality control, data processing, and data interpretation. Simplification of the sample is a key process in the direct infusion-based shotgun approach. Therefore, lipidomics can serve as a good method for food safety and quality assurance, as well as the interactions between diet and lipid metabolism. The analysis of the lipid can identify food adulteration, assure food safety, and so on. With more and more reports showing evidence that lipid metabolites can impact the diseases, lipidomics can help scientists and doctors to understand disease pathology and mechanisms of lipid-mediated disease.

In the future, it is possible to invite analytical chemists to analyze the data of lipid for a better understanding of the result and offer a more convincing suggestion, developing more novel methods to detect adulteration and evaluate food safety. With the development of analytical platforms, lipidomics can provide more information and help the regulation of food safety and quality, personalized health and nutrition, and the prevention and detection of diseases. Lipidomics is crucial to finding lipid biomarkers and can associate with other systems-based biological data to help the development of the drug for the therapy of complicated diseases.

ABBREVIATIONS

AD	Alzheimer's disease
APCI	Atmospheric Pressure Chemical Ionization
APPI	Atmospheric Pressure Photoionization
CC	Cervical Cancer
CCAD	Calcific Coronary Artery Disease
CE	Capillary Electrophoresis
CNS	Central Nervous System
CS	Calcium Score
CVD	Cardiovascular Diseases
DAG	Diacylglycerol
DAG	Diacylglycerols
DESI	Desorption Electrospray Ionization
DHA	Docosahexaenoic Acid
EPA	Eicosapentaenoic Acid
ESI	Electrospray Ionization
FA	Fatty Acid
FFA	Free Fatty Acid
FTICR	Fourier Transform Ion Cyclotron Resonance
GC	Gas Chromatography
HD	Huntington's Disease
HDL	High-Density Lipoprotein
HPLC	High-Performance Liquid Chromatography
HRMS	High-Resolution MS
HT-GC	High-Temperature Gas Chromatography
LC	Liquid Chromatography
LDL	Low Density Lipoprotein
MAG	Monoacylglycerols
MALDI	Matrix-Assisted Laser Desorption Ionization
MCI	Mild Cognitive Impairment
MDMS-SL	Multi-Dimensional MS-Based Shotgun Lipidomics

MS	Mass Spectrometry
MSI	Mass Spectrometry Imaging
NAFLD	Nonalcoholic Fatty Liver Disease
NLS	Neutral Loss Scan
NP-LC	Normal Phase Liquid Chromatography
OA	Osteoarthritis
OD	Old Dementia
PA	Palmitic Acid
PCA	Principal Component Analysis
PIS	Precursor Ion Scan
PLs	Phospholipids
PtdEth	Phosphatidylethanolamines
PUFA	Polyunsaturated Fatty Acid
Q-TOF-MS	Quadrupole Time-Of-Flight Tandem Mass Spectrometry
ROC	Receiver Operating Characteristic
RP-LC	Reverse-Phase Liquid Chromatography
SFC	Supercritical Fluid Chromatography
SIMS	Secondary Ion Mass Spectrometry
SRM/MRM	Selected/Multiple Reaction Monitoring
T1DM	Type-1 Diabetes Mellitus
TAG	Triacylglycerol
TG	Triglyceride
TLC	Thin-Layer Chromatography
TOF	Time-Of-Flight
UPLC	Ultra-Performance Liquid Chromatography
YD	Young Dementia

ACKNOWLEDGEMENT

This research has been supported by research grants from the National Key R & D Program of China (No. 2019YFD0902003)

REFERENCES

[1] Chen, H.; Tan, C.; Lin, Z.; Wu, T. Detection of melamine adulteration in milk by near-infrared spectroscopy and one-class partial least squares. *Spectrochim. Acta A Mol. Biomol. Spectrosc.,* **2017,** *173,* 832-836.
[http://dx.doi.org/10.1016/j.saa.2016.10.051] [PMID: 27816741]

[2] Ng, T-T.; So, P-K.; Zheng, B.; Yao, Z-P. Rapid screening of mixed edible oils and gutter oils by

matrix-assisted laser desorption/ionization mass spectrometry. *Anal. Chim. Acta,* **2015**, *884*, 70-76.
[http://dx.doi.org/10.1016/j.aca.2015.05.013] [PMID: 26073811]

[3]   Ickowitz, A.; Powell, B.; Rowland, D.; Jones, A.; Sunderland, T. Agricultural intensification, dietary diversity, and markets in the global food security narrative. *Glob. Food Secur.,* **2019**, *20*, 9-16.
[http://dx.doi.org/10.1016/j.gfs.2018.11.002]

[4]   Antunes, M.; Rosado, T.; Simão, A.Y.; Gonçalves, J.; Soares, S.; Barroso, M.; Gallardo, E. Liquid chromatography-mass spectrometry as a tool to identify adulteration in different food industries. In: *Food Toxicology and Forensics*; Galanakis, C.M., Ed.; Academic Press, **2021**; pp. 123-180.
[http://dx.doi.org/10.1016/B978-0-12-822360-4.00004-2]

[5]   Popa, M.E.; Mitelut, A.C.; Popa, E.E.; Stan, A.; Popa, V.I. Organic foods contribution to nutritional quality and value. *Trends Food Sci. Technol.,* **2019**, *84*, 15-18.
[http://dx.doi.org/10.1016/j.tifs.2018.01.003]

[6]   Membré, J.M.; Santillana Farakos, S.; Nauta, M. Risk-benefit analysis in food safety and nutrition. *Curr. Opin. Food Sci.,* **2021**, *39*, 76-82.
[http://dx.doi.org/10.1016/j.cofs.2020.12.009]

[7]   Orešič, M. Metabolomics, a novel tool for studies of nutrition, metabolism and lipid dysfunction. *Nutr. Metab. Cardiovasc. Dis.,* **2009**, *19*(11), 816-824.
[http://dx.doi.org/10.1016/j.numecd.2009.04.018] [PMID: 19692215]

[8]   Schmitt, S.; Cantuti Castelvetri, L.; Simons, M. Metabolism and Functions of Lipids in Myelin. *Biochimica et Biophysica Acta (BBA)- Molecular and Cell Biology of Lipids,* **2015**, *1851*, 999-1005.
[http://dx.doi.org/10.1016/j.bbalip.2014.12.016]

[9]   Proitsi, P.; Kim, M.; Whiley, L.; Simmons, A.; Sattlecker, M.; Velayudhan, L.; Lupton, M.K.; Soininen, H.; Kloszewska, I.; Mecocci, P.; Tsolaki, M.; Vellas, B.; Lovestone, S.; Powell, J.F.; Dobson, R.J.B.; Legido-Quigley, C. Association of blood lipids with Alzheimer's disease: A comprehensive lipidomics analysis. *Alzheimers Dement.,* **2017**, *13*(2), 140-151.
[http://dx.doi.org/10.1016/j.jalz.2016.08.003] [PMID: 27693183]

[10] Castelblanco, E.; Barranco, M.; Quifer, P.; Yanes, O.; Weber, R.; Ortega, E.; Alonso, N.; Mauricio, D. Lipidomic profile and subclinical carotid atherosclerosis in diabetes mellitus. *Atherosclerosis,* **2020**, *315*, e156-e157.
[http://dx.doi.org/10.1016/j.atherosclerosis.2020.10.485]

[11] Iwamoto, H.; Abe, M.; Yang, Y.; Cui, D.; Seki, T.; Nakamura, M.; Hosaka, K.; Lim, S.; Wu, J.; He, X.; Sun, X.; Lu, Y.; Zhou, Q.; Shi, W.; Torimura, T.; Nie, G.; Li, Q.; Cao, Y. Cancer lipid metabolism confers antiangiogenic drug resistance. *Cell Metab.,* **2018**, *28*(1), 104-117.e5.
[http://dx.doi.org/10.1016/j.cmet.2018.05.005] [PMID: 29861385]

[12] Wu, Z.; Bagarolo, G.I.; Thoröe-Boveleth, S.; Jankowski, J. "Lipidomics": Mass spectrometric and chemometric analyses of lipids. *Adv. Drug Deliv. Rev.,* **2020**, *159*, 294-307.
[http://dx.doi.org/10.1016/j.addr.2020.06.009] [PMID: 32553782]

[13] Kanehisa, M.; Araki, M.; Goto, S.; Hattori, M.; Hirakawa, M.; Itoh, M.; Katayama, T.; Kawashima, S.; Okuda, S.; Tokimatsu, T.; Yamanishi, Y. KEGG for linking genomes to life and the environment. *Nucleic Acids Res.,* **2008**, *36*, D480-D484.
[PMID: 18077471]

[14] Han, X.; Gross, R.W. Global analyses of cellular lipidomes directly from crude extracts of biological samples by ESI mass spectrometry: a bridge to lipidomics. *J. Lipid Res.,* **2003**, *44*(6), 1071-1079.
[http://dx.doi.org/10.1194/jlr.R300004-JLR200] [PMID: 12671038]

[15] Giles, C.; Takechi, R.; Lam, V.; Dhaliwal, S.S.; Mamo, J.C.L. Contemporary lipidomic analytics: opportunities and pitfalls. *Prog. Lipid Res.,* **2018**, *71*, 86-100.
[http://dx.doi.org/10.1016/j.plipres.2018.06.003] [PMID: 29959947]

[16] Wu, B.; Wei, F.; Xu, S.; Xie, Y.; Lv, X.; Chen, H.; Huang, F. Mass spectrometry-based lipidomics as a

powerful platform in foodomics research. *Trends Food Sci. Technol.,* **2021**, *107*, 358-376.
[http://dx.doi.org/10.1016/j.tifs.2020.10.045]

[17]   Wang, J.; Han, X. Analytical challenges of shotgun lipidomics at different resolution of measurements. *Trends Analyt. Chem.,* **2019**, *121*, 115697.
[http://dx.doi.org/10.1016/j.trac.2019.115697] [PMID: 32713986]

[18]   Luque de Castro, M.D.; Quiles-Zafra, R. Lipidomics: An omics discipline with a key role in nutrition. *Talanta,* **2020**, *219*, 121197.
[http://dx.doi.org/10.1016/j.talanta.2020.121197] [PMID: 32887107]

[19]   Wang, R.; Li, B.; Lam, S.M.; Shui, G. Integration of lipidomics and metabolomics for in-depth understanding of cellular mechanism and disease progression. *J. Genet. Genomics,* **2020**, *47*(2), 69-83.
[http://dx.doi.org/10.1016/j.jgg.2019.11.009] [PMID: 32178981]

[20]   Lee, H-C.; Yokomizo, T. Applications of mass spectrometry-based targeted and non-targeted lipidomics. *Biochem. Biophys. Res. Commun.,* **2018**, *504*(3), 576-581.
[http://dx.doi.org/10.1016/j.bbrc.2018.03.081] [PMID: 29534960]

[21]   Wei, F.; Lamichhane, S.; Orešič, M.; Hyötyläinen, T. Lipidomes in Health and Disease: Analytical Strategies and Considerations. *Trends Analyt. Chem.,* **2019**, *120*, 115664.
[http://dx.doi.org/10.1016/j.trac.2019.115664]

[22]   Shen, Q.; Wang, Y.; Gong, L.; Guo, R.; Dong, W.; Cheung, H.Y. Shotgun lipidomics strategy for fast analysis of phospholipids in fisheries waste and its potential in species differentiation. *J. Agric. Food Chem.,* **2012**, *60*(37), 9384-9393.
[http://dx.doi.org/10.1021/jf303181s] [PMID: 22946708]

[23]   Ejsing, C.S.; Moehring, T.; Bahr, U.; Duchoslav, E.; Karas, M.; Simons, K.; Shevchenko, A. Collision-induced dissociation pathways of yeast sphingolipids and their molecular profiling in total lipid extracts: a study by quadrupole TOF and linear ion trap-orbitrap mass spectrometry. *J. Mass Spectrom.,* **2006**, *41*(3), 372-389.
[http://dx.doi.org/10.1002/jms.997] [PMID: 16498600]

[24]   Han, X.; Gross, R.W. Shotgun lipidomics: electrospray ionization mass spectrometric analysis and quantitation of cellular lipidomes directly from crude extracts of biological samples. *Mass Spectrom. Rev.,* **2005**, *24*(3), 367-412.
[http://dx.doi.org/10.1002/mas.20023] [PMID: 15389848]

[25]   Han, X.; Yang, K.; Cheng, H.; Fikes, K.N.; Gross, R.W. Shotgun lipidomics of phosphoethanolamine-containing lipids in biological samples after one-step *in situ* derivatization. *J. Lipid Res.,* **2005**, *46*(7), 1548-1560.
[http://dx.doi.org/10.1194/jlr.D500007-JLR200] [PMID: 15834120]

[26]   Chao, H-C.; Chen, G-Y.; Hsu, L-C.; Liao, H-W.; Yang, S-Y.; Wang, S-Y.; Li, Y-L.; Tang, S-C.; Tseng, Y.J.; Kuo, C-H. Using precursor ion scan of 184 with liquid chromatography-electrospray ionization-tandem mass spectrometry for concentration normalization in cellular lipidomic studies. *Anal. Chim. Acta,* **2017**, *971*, 68-77.
[http://dx.doi.org/10.1016/j.aca.2017.03.033] [PMID: 28456285]

[27]   Yeo, J.; Parrish, C.C. (Tag) Profiles and Their Contents in Salmon Muscle Tissue Using Esi-Ms/Ms Spectrometry with Multiple Neutral Loss Scans. *Food Chem.,* **2020**, *324*, 126816.
[http://dx.doi.org/10.1016/j.foodchem.2020.126816] [PMID: 32344337]

[28]   Tans, R.; Bande, R.; van Rooij, A.; Molloy, B.J.; Stienstra, R.; Tack, C.J.; Wevers, R.A.; Wessels, H.J.C.T.; Gloerich, J.; van Gool, A.J. Evaluation of cyclooxygenase oxylipins as potential biomarker for obesity-associated adipose tissue inflammation and type 2 diabetes using targeted multiple reaction monitoring mass spectrometry. *Prostaglandins Leukot. Essent. Fatty Acids,* **2020**, *160*, 102157.
[http://dx.doi.org/10.1016/j.plefa.2020.102157] [PMID: 32629236]

[29]   Yang, K.; Han, X. Lipidomics: techniques, applications, and outcomes related to biomedical sciences. *Trends Biochem. Sci.,* **2016**, *41*(11), 954-969.

[http://dx.doi.org/10.1016/j.tibs.2016.08.010] [PMID: 27663237]

[30]   Fu, T.; Knittelfelder, O.; Geffard, O.; Clément, Y.; Testet, E.; Elie, N.; Touboul, D.; Abbaci, K.; Shevchenko, A.; Lemoine, J.; Chaumot, A.; Salvador, A.; Degli-Esposti, D.; Ayciriex, S. Shotgun lipidomics and mass spectrometry imaging unveil diversity and dynamics in *Gammarus fossarum* lipid composition. *iScience,* **2021**, *24*(2), 102115.
[http://dx.doi.org/10.1016/j.isci.2021.102115] [PMID: 33615205]

[31]   Brügger, B.; Erben, G.; Sandhoff, R.; Wieland, F.T.; Lehmann, W.D. Quantitative analysis of biological membrane lipids at the low picomole level by nano-electrospray ionization tandem mass spectrometry. *Proc. Natl. Acad. Sci. USA,* **1997**, *94*(6), 2339-2344.
[http://dx.doi.org/10.1073/pnas.94.6.2339] [PMID: 9122196]

[32]   Liebisch, G.; Lieser, B.; Rathenberg, J.; Drobnik, W.; Schmitz, G. High-throughput quantification of phosphatidylcholine and sphingomyelin by electrospray ionization tandem mass spectrometry coupled with isotope correction algorithm. *Biochim. Biophys. Acta,* **2004**, *1686*(1-2), 108-117.
[http://dx.doi.org/10.1016/j.bbalip.2004.09.003] [PMID: 15522827]

[33]   Welti, R.; Li, W.; Li, M.; Sang, Y.; Biesiada, H.; Zhou, H.E.; Rajashekar, C.B.; Williams, T.D.; Wang, X. Profiling membrane lipids in plant stress responses. Role of phospholipase D alpha in freezing-induced lipid changes in Arabidopsis. *J. Biol. Chem.,* **2002**, *277*(35), 31994-32002.
[http://dx.doi.org/10.1074/jbc.M205375200] [PMID: 12077151]

[34]   Xu, X.; Luo, Z.; He, Y.; Shan, J.; Guo, J.; Li, J. Application of untargeted lipidomics based on UHPLC-high resolution tandem MS analysis to profile the lipid metabolic disturbances in the heart of diabetic cardiomyopathy mice. *J. Pharm. Biomed. Anal.,* **2020**, *190*, 113525.
[http://dx.doi.org/10.1016/j.jpba.2020.113525] [PMID: 32827999]

[35]   Schwudke, D.; Oegema, J.; Burton, L.; Entchev, E.; Hannich, J.T.; Ejsing, C.S.; Kurzchalia, T.; Shevchenko, A. Lipid profiling by multiple precursor and neutral loss scanning driven by the data-dependent acquisition. *Anal. Chem.,* **2006**, *78*(2), 585-595.
[http://dx.doi.org/10.1021/ac051605m] [PMID: 16408944]

[36]   Wang, M.; Wang, C.; Han, R.H.; Han, X. Novel advances in shotgun lipidomics for biology and medicine. *Prog. Lipid Res.,* **2016**, *61*, 83-108.
[http://dx.doi.org/10.1016/j.plipres.2015.12.002] [PMID: 26703190]

[37]   Jiang, X.; Cheng, H.; Yang, K.; Gross, R.W.; Han, X. Alkaline methanolysis of lipid extracts extends shotgun lipidomics analyses to the low-abundance regime of cellular sphingolipids. *Anal. Biochem.,* **2007**, *371*(2), 135-145.
[http://dx.doi.org/10.1016/j.ab.2007.08.019] [PMID: 17920553]

[38]   Jin, R.; Li, L.; Feng, J.; Dai, Z.; Huang, Y-W.; Shen, Q. Zwitterionic hydrophilic interaction solid-phase extraction and multi-dimensional mass spectrometry for shotgun lipidomic study of Hypophthalmichthys nobilis. *Food Chem.,* **2017**, *216*, 347-354.
[http://dx.doi.org/10.1016/j.foodchem.2016.08.074] [PMID: 27596430]

[39]   Wang, C.; Wang, M.; Han, X. Applications of mass spectrometry for cellular lipid analysis. *Mol. Biosyst.,* **2015**, *11*(3), 698-713.
[http://dx.doi.org/10.1039/C4MB00586D] [PMID: 25598407]

[40]   Paglia, G.; Kliman, M.; Claude, E.; Geromanos, S.; Astarita, G. Applications of ion-mobility mass spectrometry for lipid analysis. *Anal. Bioanal. Chem.,* **2015**, *407*(17), 4995-5007.
[http://dx.doi.org/10.1007/s00216-015-8664-8] [PMID: 25893801]

[41]   Lintonen, T.P.; Baker, P.R.; Suoniemi, M.; Ubhi, B.K.; Koistinen, K.M.; Duchoslav, E.; Campbell, J.L.; Ekroos, K. Differential mobility spectrometry-driven shotgun lipidomics. *Anal. Chem.,* **2014**, *86*(19), 9662-9669.
[http://dx.doi.org/10.1021/ac5021744] [PMID: 25160652]

[42]   Gross, R.W. The evolution of lipidomics through space and time. *Biochim. Biophys. Acta Mol. Cell Biol. Lipids,* **2017**, *1862*(8), 731-739.

[http://dx.doi.org/10.1016/j.bbalip.2017.04.006] [PMID: 28457845]

[43]   Cajka, T.; Fiehn, O. Comprehensive analysis of lipids in biological systems by liquid chromatography-mass spectrometry. *Trends Analyt. Chem.,* **2014**, *61*, 192-206.
[http://dx.doi.org/10.1016/j.trac.2014.04.017] [PMID: 25309011]

[44]   Chen, G.Y.; Chiu, H.H.; Lin, S.W.; Tseng, Y.J.; Tsai, S.J.; Kuo, C.H. Development and application of a comparative fatty acid analysis method to investigate voriconazole-induced hepatotoxicity. *Clin. Chim. Acta,* **2015**, *438*, 126-134.
[http://dx.doi.org/10.1016/j.cca.2014.08.013] [PMID: 25150729]

[45]   Lísa, M.; Cífková, E.; Khalikova, M.; Ovčačíková, M.; Holčapek, M. Lipidomic analysis of biological samples: Comparison of liquid chromatography, supercritical fluid chromatography and direct infusion mass spectrometry methods. *J. Chromatogr. A,* **2017**, *1525*, 96-108.
[http://dx.doi.org/10.1016/j.chroma.2017.10.022] [PMID: 29037587]

[46]   Gao, F.; Dong, J.; Li, W.; Wang, T.; Liao, J.; Liao, Y.; Liu, H. Separation of phospholipids by capillary zone electrophoresis with indirect ultraviolet detection. *J. Chromatogr. A,* **2006**, *1130*(2), 259-264.
[http://dx.doi.org/10.1016/j.chroma.2006.03.070] [PMID: 16620855]

[47]   Bowman, A.P.; Heeren, R.M.A.; Ellis, S.R. Advances in mass spectrometry imaging enabling observation of localised lipid biochemistry within tissues. *Trends Analyt. Chem.,* **2019**, *120*, 115197.
[http://dx.doi.org/10.1016/j.trac.2018.07.012]

[48]   Ewing, A.G.; Lanekoff, I.; Kurczy, M.; Sjövall, P. Relative Quantification of Exogenous Phospholipids Regulating Exocytosis by Tof Sims of Hydrated Single Cells. *Biophys. J.,* **2011**, *100*, 37a-38a.
[http://dx.doi.org/10.1016/j.bpj.2010.12.408]

[49]   Li, M.; Yang, L.; Bai, Y.; Liu, H. Analytical methods in lipidomics and their applications. *Anal. Chem.,* **2014**, *86*(1), 161-175.
[http://dx.doi.org/10.1021/ac403554h] [PMID: 24215393]

[50]   Collino, S.; Martin, F-P.; Moco, S. *Metabonomics and Gut Microbiota in Nutrition and Disease*; Kochhar, S.; Martin, F-P., Eds.; Springer London: London, **2015**, pp. 25-44.
[http://dx.doi.org/10.1007/978-1-4471-6539-2_2]

[51]   Takis, P.G.; Ghini, V.; Tenori, L.; Turano, P.; Luchinat, C. Uniqueness of the Nmr Approach to Metabolomics. *Trends Analyt. Chem.,* **2019**, *120*, 115300.
[http://dx.doi.org/10.1016/j.trac.2018.10.036]

[52]   Li, J.; Vosegaard, T.; Guo, Z. Applications of nuclear magnetic resonance in lipid analyses: An emerging powerful tool for lipidomics studies. *Prog. Lipid Res.,* **2017**, *68*, 37-56.
[http://dx.doi.org/10.1016/j.plipres.2017.09.003] [PMID: 28911967]

[53]   Merrill, A.H., Jr; Sullards, M.C.; Allegood, J.C.; Kelly, S.; Wang, E. Sphingolipidomics: high-throughput, structure-specific, and quantitative analysis of sphingolipids by liquid chromatography tandem mass spectrometry. *Methods,* **2005**, *36*(2), 207-224.
[http://dx.doi.org/10.1016/j.ymeth.2005.01.009] [PMID: 15894491]

[54]   Wu, T.; Guo, H.; Lu, Z.; Zhang, T.; Zhao, R.; Tao, N.; Wang, X.; Zhong, J. Reliability of LipidSearch software identification and its application to assess the effect of dry salting on the long-chain free fatty acid profile of tilapia muscles. *Food Res. Int.,* **2020**, *138*(Pt B), 109791.
[http://dx.doi.org/10.1016/j.foodres.2020.109791] [PMID: 33288177]

[55]   Shi, C.; Guo, H.; Wu, T.; Tao, N.; Wang, X.; Zhong, J. Effect of three types of thermal processing methods on the lipidomics profile of tilapia fillets by UPLC-Q-Extractive Orbitrap mass spectrometry. *Food Chem.,* **2019**, *298*, 125029.
[http://dx.doi.org/10.1016/j.foodchem.2019.125029] [PMID: 31260974]

[56]   Yetukuri, L.; Katajamaa, M.; Medina-Gomez, G.; Seppänen-Laakso, T.; Vidal-Puig, A.; Orešič, M.

Bioinformatics strategies for lipidomics analysis: characterization of obesity related hepatic steatosis. *BMC Syst. Biol.,* **2007**, *1,* 12.
[http://dx.doi.org/10.1186/1752-0509-1-12] [PMID: 17408502]

[57]   Xia, J.; Psychogios, N.; Young, N.; Wishart, D.S. MetaboAnalyst: a web server for metabolomic data analysis and interpretation. *Nucleic Acids Res.,* **2009**, *37*(Web Server issue), W652-60.
[http://dx.doi.org/10.1093/nar/gkp356] [PMID: 19429898]

[58]   Chong, J.; Soufan, O.; Li, C.; Caraus, I.; Li, S.; Bourque, G.; Wishart, D.S.; Xia, J. MetaboAnalyst 4.0: towards more transparent and integrative metabolomics analysis. *Nucleic Acids Res.,* **2018**, *46*(W1), W486-W494.
[http://dx.doi.org/10.1093/nar/gky310] [PMID: 29762782]

[59]   Indelicato, S.; Bongiorno, D.; Pitonzo, R.; Di Stefano, V.; Calabrese, V.; Indelicato, S.; Avellone, G. Triacylglycerols in edible oils: Determination, characterization, quantitation, chemometric approach and evaluation of adulterations. *J. Chromatogr. A,* **2017**, *1515,* 1-16.
[http://dx.doi.org/10.1016/j.chroma.2017.08.002] [PMID: 28801042]

[60]   Hou, X.; Wang, G.; Su, G.; Wang, X.; Nie, S. Rapid identification of edible oil species using supervised support vector machine based on low-field nuclear magnetic resonance relaxation features. *Food Chem.,* **2019**, *280,* 139-145.
[http://dx.doi.org/10.1016/j.foodchem.2018.12.031] [PMID: 30642479]

[61]   Ng, T-T.; Li, S.; Ng, C.C.A.; So, P-K.; Wong, T-F.; Li, Z-Y.; Chan, S-T.; Yao, Z-P. Establishment of a spectral database for classification of edible oils using matrix-assisted laser desorption/ionization mass spectrometry. *Food Chem.,* **2018**, *252,* 335-342.
[http://dx.doi.org/10.1016/j.foodchem.2018.01.125] [PMID: 29478551]

[62]   Hrbek, V.; Vaclavik, L.; Elich, O.; Hajslova, J. Authentication of Milk and Milk-Based Foods by Direct Analysis in Real Time Ionization-High Resolution Mass Spectrometry (Dart-Hrms) Technique: A Critical Assessment. *Food Control,* **2014**, *36,* 138-145.
[http://dx.doi.org/10.1016/j.foodcont.2013.08.003]

[63]   Rebechi, S.R.; Vélez, M.A.; Vaira, S.; Perotti, M.C. Adulteration of Argentinean milk fats with animal fats: Detection by fatty acids analysis and multivariate regression techniques. *Food Chem.,* **2016**, *192,* 1025-1032.
[http://dx.doi.org/10.1016/j.foodchem.2015.07.107] [PMID: 26304443]

[64]   Trocino, A.; Xiccato, G.; Majolini, D.; Tazzoli, M.; Bertotto, D.; Pascoli, F.; Palazzi, R. Assessing the Quality of Organic and Conventionally-Farmed European Sea Bass (*Dicentrarchus Labrax*). *Food Chem.,* **2012**, *131,* 427-433.
[http://dx.doi.org/10.1016/j.foodchem.2011.08.082]

[65]   Fiorino, G.M.; Losito, I.; De Angelis, E.; Arlorio, M.; Logrieco, A.F.; Monaci, L. Assessing fish authenticity by direct analysis in real time-high resolution mass spectrometry and multivariate analysis: discrimination between wild-type and farmed salmon. *Food Res. Int.,* **2019**, *116,* 1258-1265.
[http://dx.doi.org/10.1016/j.foodres.2018.10.013] [PMID: 30716913]

[66]   Peters, B.S.; Wierzbicki, A.S.; Moyle, G.; Nair, D.; Brockmeyer, N. The effect of a 12-week course of omega-3 polyunsaturated fatty acids on lipid parameters in hypertriglyceridemic adult HIV-infected patients undergoing HAART: a randomized, placebo-controlled pilot trial. *Clin. Ther.,* **2012**, *34*(1), 67-76.
[http://dx.doi.org/10.1016/j.clinthera.2011.12.001] [PMID: 22212377]

[67]   Calder, P.C. The role of marine omega-3 (n-3) fatty acids in inflammatory processes, atherosclerosis and plaque stability. *Mol. Nutr. Food Res.,* **2012**, *56*(7), 1073-1080.
[http://dx.doi.org/10.1002/mnfr.201100710] [PMID: 22760980]

[68]   Lamaziere, A.; Wolf, C.; Barbe, U.; Bausero, P.; Visioli, F. Lipidomics of hepatic lipogenesis inhibition by omega 3 fatty acids. *Prostaglandins Leukot. Essent. Fatty Acids,* **2013**, *88*(2), 149-154.
[http://dx.doi.org/10.1016/j.plefa.2012.12.001] [PMID: 23313470]

[69]   Kunisawa, J.; Kiyono, H. Sphingolipids and epoxidized lipid metabolites in the control of gut immunosurveillance and allergy. *Front. Nutr.,* **2016**, *3*, 3.
[http://dx.doi.org/10.3389/fnut.2016.00003] [PMID: 26858949]

[70]   Heilbronn, L.K.; Coster, A.C.; Campbell, L.V.; Greenfield, J.R.; Lange, K.; Christopher, M.J.; Meikle, P.J.; Samocha-Bonet, D. The effect of short-term overfeeding on serum lipids in healthy humans. *Obesity (Silver Spring),* **2013**, *21*(12), E649-E659.
[http://dx.doi.org/10.1002/oby.20508] [PMID: 23640727]

[71]   Rey, C.; Delpech, J.C.; Madore, C.; Nadjar, A.; Greenhalgh, A.D.; Amadieu, C.; Aubert, A.; Pallet, V.; Vaysse, C.; Layé, S.; Joffre, C. Dietary n-3 long chain PUFA supplementation promotes a pro-resolving oxylipin profile in the brain. *Brain Behav. Immun.,* **2019**, *76*, 17-27.
[http://dx.doi.org/10.1016/j.bbi.2018.07.025] [PMID: 30086401]

[72]   Astarita, G.; Piomelli, D. Towards a whole-body systems [multi-organ] lipidomics in Alzheimer's disease. *Prostaglandins Leukot. Essent. Fatty Acids,* **2011**, *85*(5), 197-203. [PLEFA].
[http://dx.doi.org/10.1016/j.plefa.2011.04.021] [PMID: 21543199]

[73]   Proitsi, P.; Kim, M.; Whiley, L.; Pritchard, M.; Leung, R.; Soininen, H.; Kloszewska, I.; Mecocci, P.; Tsolaki, M.; Vellas, B.; Sham, P.; Lovestone, S.; Powell, J.F.; Dobson, R.J.B.; Legido-Quigley, C. Plasma lipidomics analysis finds long chain cholesteryl esters to be associated with Alzheimer's disease. *Transl. Psychiatry,* **2015**, *5*, e494-e494.
[http://dx.doi.org/10.1038/tp.2014.127] [PMID: 25585166]

[74]   Wood, P.L.; Barnette, B.L.; Kaye, J.A.; Quinn, J.F.; Woltjer, R.L. Non-targeted lipidomics of CSF and frontal cortex grey and white matter in control, mild cognitive impairment, and Alzheimer's disease subjects. *Acta Neuropsychiatr.,* **2015**, *27*(5), 270-278.
[http://dx.doi.org/10.1017/neu.2015.18] [PMID: 25858158]

[75]   Vodicka, P.; Mo, S.; Tousley, A.; Green, K.M.; Sapp, E.; Iuliano, M.; Sadri-Vakili, G.; Shaffer, S.A.; Aronin, N.; DiFiglia, M.; Kegel-Gleason, K.B. Mass Spectrometry Analysis of Wild-Type and Knock-in Q140/Q140 Huntington's Disease Mouse Brains Reveals Changes in Glycerophospholipids Including Alterations in Phosphatidic Acid and Lyso-Phosphatidic Acid. *J. Huntingtons Dis.,* **2015**, *4*(2), 187-201.
[http://dx.doi.org/10.3233/JHD-150149] [PMID: 26397899]

[76]   Ferreira, H.B.; Neves, B.; Guerra, I.M.; Moreira, A.; Melo, T.; Paiva, A.; Domingues, M.R. An overview of lipidomic analysis in different human matrices of multiple sclerosis. *Mult. Scler. Relat. Disord.,* **2020**, *44*, 102189.
[http://dx.doi.org/10.1016/j.msard.2020.102189] [PMID: 32516740]

[77]   Meikle, P.J.; Wong, G.; Barlow, C.K.; Kingwell, B.A. Lipidomics: potential role in risk prediction and therapeutic monitoring for diabetes and cardiovascular disease. *Pharmacol. Ther.,* **2014**, *143*(1), 12-23.
[http://dx.doi.org/10.1016/j.pharmthera.2014.02.001] [PMID: 24509229]

[78]   Hinterwirth, H.; Stegemann, C.; Mayr, M. Lipidomics: quest for molecular lipid biomarkers in cardiovascular disease. *Circ. Cardiovasc. Genet.,* **2014**, *7*(6), 941-954.
[http://dx.doi.org/10.1161/CIRCGENETICS.114.000550] [PMID: 25516624]

[79]   Zhang, J.; Liang, S.; Ning, R.; Jiang, J.; Zhang, J.; Shen, H.; Chen, R.; Duan, J.; Sun, Z. $PM_{2.5}$-induced inflammation and lipidome alteration associated with the development of atherosclerosis based on a targeted lipidomic analysis. *Environ. Int.,* **2020**, *136*, 105444.
[http://dx.doi.org/10.1016/j.envint.2019.105444] [PMID: 31935561]

[80]   Yan, Y.; Du, Z.; Chen, C.; Li, J.; Xiong, X.; Zhang, Y.; Jiang, H. Lysophospholipid profiles of apolipoprotein E-deficient mice reveal potential lipid biomarkers associated with atherosclerosis progression using validated UPLC-QTRAP-MS/MS-based lipidomics approach. *J. Pharm. Biomed. Anal.,* **2019**, *171*, 148-157.
[http://dx.doi.org/10.1016/j.jpba.2019.03.062] [PMID: 30999225]

[81]   Sobczak, A.I.S.; Pitt, S.J.; Smith, T.K.; Ajjan, R.A.; Stewart, A.J. Lipidomic profiling of plasma free fatty acids in type-1 diabetes highlights specific changes in lipid metabolism. *Biochim. Biophys. Acta Mol. Cell Biol. Lipids,* **2021**, *1866*(1), 158823.
[http://dx.doi.org/10.1016/j.bbalip.2020.158823] [PMID: 33010452]

[82]   Cheng, F.; Wen, Z.; Feng, X.; Wang, X.; Chen, Y. A serum lipidomic strategy revealed potential lipid biomarkers for early-stage cervical cancer. *Life Sci.,* **2020**, *260*, 118489.
[http://dx.doi.org/10.1016/j.lfs.2020.118489] [PMID: 32976882]

[83]   Ahmedin J., Rebecca S., Elizabeth W., Yongping H., Jiaquan X., Taylor M., Michael J. Thun Cancer Statistics. *CA Cancer J. Clin.,* **2008**, *20*, 10-23.

[84]   Bedia, C.; Dalmau, N.; Jaumot, J.; Tauler, R. Phenotypic malignant changes and untargeted lipidomic analysis of long-term exposed prostate cancer cells to endocrine disruptors. *Environ. Res.,* **2015**, *140*, 18-31.
[http://dx.doi.org/10.1016/j.envres.2015.03.014] [PMID: 25817993]

[85]   Rocha, B.; Cillero-Pastor, B.; Ruiz-Romero, C.; Paine, M.R.L.; Cañete, J.D.; Heeren, R.M.A.; Blanco, F.J. Identification of a distinct lipidomic profile in the osteoarthritic synovial membrane by mass spectrometry imaging. *Osteoarthritis Cartilage,* **2021**, *29*(5), 750-761.
[http://dx.doi.org/10.1016/j.joca.2020.12.025] [PMID: 33582239]

[86]   Sun, Q.; Yu, X.; Peng, C.; Liu, N.; Chen, W.; Xu, H.; Wei, H.; Fang, K.; Dong, Z.; Fu, C.; Xu, Y.; Lu, W. Activation of SREBP-1c alters lipogenesis and promotes tumor growth and metastasis in gastric cancer. *Biomed. Pharmacother.,* **2020**, *128*, 110274.
[http://dx.doi.org/10.1016/j.biopha.2020.110274] [PMID: 32464305]

CHAPTER 3

Recent Advances in the Analysis of Herbicides and their Transformation Products in Environmental Samples

Pervinder Kaur[1,*], **Harshdeep Kaur**[2] and **Makhan Singh Bhullar**[1]

[1] *Department of Agronomy, Punjab Agricultural University, Ludhiana, Punjab, India*

[2] *Department of Chemistry, Punjab Agricultural University, Ludhiana, Punjab, India*

Abstract: Herbicide residues in crop, soil and contamination of groundwater have become a worldwide concern in recent decades as their presence at low concentrations entails an unacceptable risk to human health and non-target organism. The magnitude of exposure and concentration at a particular time may trigger bioaccumulation and bio magnifications of herbicide residues and their degraded products, causing mutagenic, carcinogenic and teratogenic effects on humans, flora and fauna and microbiological living system. These atrocious circumstances have raised concern about their presence in environmental compartments and necessitate the continuous monitoring of herbicide residues in various matrices. However, determining the herbicide residues in the soil and crop is challenging because of the very low concentration of analyte, complicated sample matrices and low maximum residue limit (MRL) imposed by the regulatory agencies. The detection limits imposed by environment quality legislation can only be achieved by using appropriate sample preparation techniques, which comprise isolation and concentration of the analytes with nominal matrix interference, thus allowing its facile detection and quantification through instrumental analysis. In recent years, the requirements for separation and pre-concentration procedures have undergone numerous changes, and various sample preparation methods have been used. The final step in the analytical process involves the identification and quantification of the herbicide residues using suitable instrumentation, and over the years, herbicides have been determined by spectrophotometric, chromatographic, electrochemical, electrophoretic, hyphenated and biosensors. This book chapter provides a comprehensive overview of the novelties and the advantages of different techniques employed for the detection of herbicides and their transformation products in environmental samples.

Keywords: Biosensors, Chromatographic Methods, Electrochemical, Herbicides, Hyphenated, Maximum Residue Limit, Microextraction, Pretreatment, Quantification, Transformation Products.

* **Corresponding author Pervinder Kaur**: Department of Agronomy, Punjab Agricultural University, Ludhiana, Punjab, India; E-mail: pervi_7@yahoo.co.in

Sibel A. Ozkan (Ed.)

INTRODUCTION

Weed management is essential for agricultural production and will play an important role to meet future food production requirements. Over the years, herbicides have emerged as an effective method for controlling weeds. Herbicides use has been increasing throughout the globe due to increasing labour cost, choice of application of herbicides, quick weed control in crop and non-crop areas, *etc.* Significant amounts are also used in the lawns, parks, golf courses, forestry, industry and for maintenance of rights-of-way for pipelines, power-lines and highways. Herbicides are among the largest growing segments, accounting for 60% (Fig. **1a**) of total crop protection globally, and their use is projected to increase by 15-20% per year. Of the total herbicide consumption, rice and wheat account for a major share of 25.0 and 20.0% in the world market. In maize, total herbicide consumption is 15%, and vegetables and oilseed pulses account for 10-15% of the herbicide consumption. Fruits, cotton, sugarcane and other crops involve 4-6% herbicide consumption (Fig. **1b**). Table **1** enlists some of the widely used herbicide classes for control of weeds in different crops.

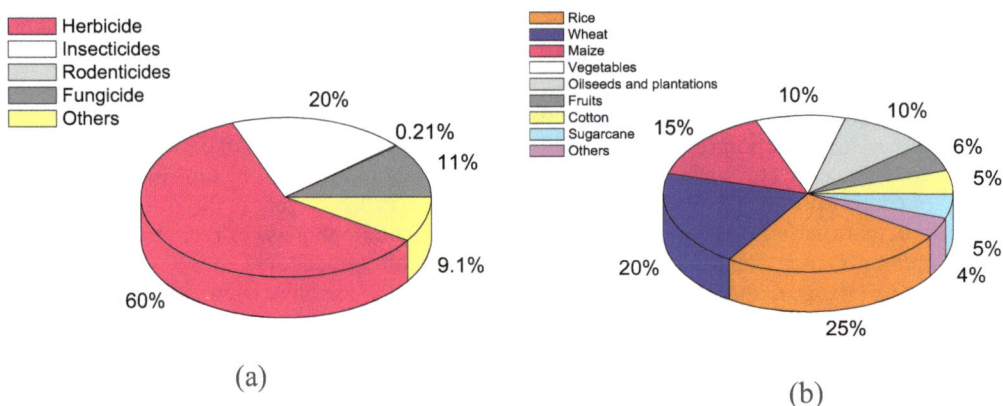

Fig. (1). (a) Total pesticide consumption in the world **(b)** Total herbicide consumption in different crops [1].

Though herbicides help in the protection of crops from weeds and are vital for higher productivity at a lower cost but their indiscriminate and non-judicious usage can ultimately cause health hazards because of bioaccumulation and biomagnifications of herbicide residues. Not only herbicides but their transformation products (Table **1**) can also pose a residual problem and are thus required to be considered while taking the environmental risk estimation (Table **1**).

Increased concern about the presence of herbicides in environmental compartments necessitates the continuous monitoring of herbicide residues and their transformation products in various matrices. International agencies, such as European Union (EU), CODEX and USEPA have decided on maximum residue limits (MRLs) (Table **2**) in order to keep a check on herbicide residues in various food commodities and keep them within safe limits. It is mandatory for herbicides to meet the criteria for food safety regulation to consider them safe for application.

Table 1. Transformation products (TPs) of herbicides with their molecular formula.

Herbicide	Transformation Product/Metabolites	Matrix	References
Penoxsulam	2-amino-TP, 5-OH-XDE-638, BSTCA, BSTCA-methyl and BST	Soil	[2]
Bispyribac sodium	Sodium 2-(4,6-dimethoxy-2- pyrimidinyl)oxy)-6-((4-hydro-y-6-methoxy-2-pyrimidinyl)oxy benzoate and sodium 2-((4,--dimethoxy-2-pyrimidinyl)oxy) 6- hydroxy benzoate	Soil	[3]
Linuron	-N(OCH)CH$_2$OH, -NHCH$_3$, and -NHOCH$_3$	Plant	[4]
2, 4-D	2,4-dichlorophenol (2,4-DCP), 2-chlorophenol (2-CP), pbenzoquinone (PBQ), 2-chlorohydroquinone (2-CHQ) and 4-chloro1,3-benzenediol (4-Chlororesorcinol, 4-CR)	Water	[5]
Pendimethalin	N-(1-ethyl1-propyl)-3,4 dicarboxy-2,6-dinitrobenzenamine-N-oxide, N-(1-ethylpropyl)-3,4 dimethoxy-2,6-dinitrobenzenamine and benzimadazole-7-carboxyaldehyde	Soil	[6]
Atrazine	Desethylatrazine (DEA) and Deisopropylatrazine (DIA)	Water	[7]
Simazine	2-hydroxysimazine	Water	[8]
Butachlor	N-(butoxymethyl)-N-(2-chloroethyl)-2,6-diethylaniline, (N-(butoxymethyl)- 2-chloro-N-(2-ethylphenyl) acetamide, N-(butoxymethyl)-2,6-diethyl-N-propylaniline, 2-chloro-N-(-,6-diethylphenyl) acetamide and 2,6-diethylaniline	Soil	[9]
Pretilachlor	N-ethyl-2-chloro-2′,6′-diethylacetanilide, 2-hydroxy2′,6-diethylacetanilide, 2,6-diethyl-N- (propyloxyethyl)acetanilide and 2,6-diethyl-N-(propyloxyethyl)aniline	Soil	[10]
Quizalofop	6-chloroquinoxalin-2-ol), (R)-2-(4-hydroxyphenoxy)propionic acid and 2,3-dihydroxyquinoxaline	Soils	[11]
Cyhalofop- butyl	Cyhalofop acid	Soil	[12]
Clodinafop -propargyl	Acid derivative-clodinafop	Wheat	[13]
Imazamox	5-(methoxymethyl)-2,3-pyridinecarboxylic acid and 2-carbamoyl-5-(methoxymethyl)-3-pyridinecarboxylic acid	Water	[14]
Imazethapyr	2,3-pyridinecarboxylic acid and 7- hydroxy-furo[3,4-b]pyridine-5(-H)-one	Water	[14]
Triflusulfuron	2-amino-4-(dimethylamino)-6-(2,2,2-trifluoroethoxy)-1,3,5-triazine (2) and 6-methyl-2- methylcarboxylate benzene sulfonamide	Soil	[15]
Prosulfuron	phenyl sulfonamide, desmethyl prosulfuron and amino triazine	Soil	[16]

(Table 1) cont.....

Herbicide	Transformation Product/Metabolites	Matrix	References
Foransulam	N-(2,6-difluorophenyl)-5-aminosulfonyl-1H-1,2,4-triazole-3-carboxylic acid, N-(2,6-difluorophenyl)-8-fluoro-5-hydroxy [1, 2, 4]triazolo[1,5-c]pyrimidine-2- sulfonamide and m 5-(aminosulfonyl)- 1H-1,2,--triazole-3-carboxylic acid	Soil	[17]
Nicosulfuron	2-aminosulfonyl-N, N-dimethylnicotinamide, 4, 6-dihydroxypyrimidine, 2-amino-4, 6-dimethoxypyrimidine and 2-(1-(4,6-dimethoxy-pyrim-din-2-yl)-ureido)-N,N-dimethyl-nicotinamide	Soil	[18]
Chlorsulfuron	5-hydroxy-chlorobenzenesulfonamide and 2-chloro-5-hydroxy-N-[(4-methoxy-6-methyl- 1,3,5-triazi--2yl)aminocarbonyl]benzenesulfonamide	Wheat	[19]
Sulfosulfuron	1-(2-ethylsulfonylimidazo[1,2-a]- pyridin-3-yl-3--4,6-dimethoxypyrimidin-2-yl) amine,o 1-(2- ethylsulfonylimidazo[1,2-a] pyridin)-3-sulfonamide and 4,6-dimethoxy-2-aminopyrimidine	Soil	[20]
Chlorimuron-ethyl	N-(4-methoxy-6-chloropyrimidin-2-yl)methyl urea, 4-methoxy-6-chlro-2- aminopyrimidine, o-benzoic sulfimide (saccharin) and ethyl 2-aminosulfonylbenzoate	Soybean	[21]
Clomazone	5-ketoclomazone	Rice	[22]
Metolachlor	4-(2-ethyl-6-methylphenyl)-5-methyl-3-morpholinone and N-(2-ethy--6-methylphenyl)2-hydroxy-n-(2-methylethyl)-acetamide	Soil	[23]
Mesotrione	[2-amino-4(methylsulfonyl) benzoic acid]) and [4-(methylsulfonyl)-2-nitrobenzoic acid])	Soil	[24]
Carfentrazone ethyl	carfentrazone-chloropropionic acid, carfentrazone-cinnamic acid, carfentrazone-propionic acid, carfentrazone-benzoic acid and 3-(hydroxymethyl)carfentrazone-benzoic acid	Soil	[25]
Flumetsulam	[N-(2,6-dimethylphenyl)-5,7-dimethyl [1, 2, 4]triazolo[1,5-a]pyrimidine-2-sulfonamide] and [N-(2-methylpheny-)-5,7-dimethyl [1, 2, 4] triazolo[1,5-a]pyrimidine-2-sulfonamide]	Soil	[26]

Table 2. Physico-chemical characteristics of herbicides and their maximum residue limit in various recommended crops.

Class	Herbicide	Recommended in Crops	MRL Values
Triazines	Atrazine	Sorghum, sugarcane and maize	0.05
	Simazine	Nuts, cherries and fruit crops	0.05
	Propazine	Sorghum	0.01-0.02
	Ametryn	Corn, sugarcane and fruit crops	0.04
	Terbutryn	Sorghum and wheat	0.01

(Table 2) cont.....

Class	Herbicide	Recommended in Crops	MRL Values
Sulfonylureas	Sulfosulfuron	Wheat, potato and tomato	0.01-0.05
	Metsulfuron methyl	Wheat, rice and sugarcane	0.01-0.02
	Thifensulfuron	Wheat, barley and oats	0.01-0.05
	Ethametsulfuron	Wheat, rice and sugarcane	0.01
	Mesosulfuron methyl	Wheat and rye	0.01-0.05
	Iodosulfuron methyl	Wheat	0.01-0.05
	Primisulfuron	Corn and sorghum	0.01
	Chlorsulfuron	Wheat, barley, oats and cereal rye	0.01
	Chlorimuron	Paddy and soyabean	0.01
	Ethoxysulfuron	Rice	0.01
Dinitroaniline	Fluchloralin	Beets, sorghum, spinach and oats	0.01
	Pendimethalin	Wheat, cotton, corn and soyabeans	0.05
	Trifluralin	Soyabeans and cotton	0.01
Chloroacetanilide	Butachlor	Rice and barley	0.01
	Pretilachlor	Paddy	0.01
	Metolachlor	Corn, cotton and peanuts	0.02-0.05
	Alachlor	Corn, sorghum and soyabeans	0.01
Aryloxyphenoxy propionates	Fenoxaprop-ethyl	Sugarcane, cotton and rice	0.01
	Metamifop	Rice	0.01
	Quizalofop	Sugarbeet, sunflower, potatoes and oilseed rape	0.01
	Cyhalofop- butyl	Rice	0.01-0.02
	Clodinafop -propargyl	Wheat	0.01
Imidazolinones	Imazamox	Beans and peas	0.05
	Imazethapyr	Soyabean	0.01-0.15
	Imazapyr	Pulses, cereals and oilseeds	0.01-0.08
Triazolopyrimidine	Sulfonanilide	Rice and wheat	0.01
	Cloransulam	Beans	0.01
	Flumetsulam	Lentils, maize and peanuts	0.01
	Penoxsulam	Rice	0.01
Isoxazolidinones	Clomazone	Soyabean, maize and sugarcane	0.05
Diphenylethers	Oxyfluorfen	Rice, cotton and cabbage	0.01-0.05
	Aclonifen	Beans	0.01-0.03

(Table 2) cont.....

Class	Herbicide	Recommended in Crops	MRL Values
Triketone	Mesotrione	Corn and maize	0.01-0.05
	Tembotrione	Corn	0.02-0.05
Acetamide	Pyroxasulfone	Corn and soy	0.01
Carboxamide	Diflufenican	Wheat, barley and rye	0.2
Aryl triazolinones	Carfentrazone ethyl	Potatoes, barley, wheat and oats	0.01-0.05
Phenylpyrazolines	Pinoxaden	Wheat and barley	0.01
Pyrimidinyloxy benzoic acid	Bispyribac sodium	Rice	0.01-0.02
Phenylurea	Isoproturon	Wheat, sugarcane and cotton	0.01
	Linuron	Wheat, carrot and fruit crops	0.01
Phenoxyacetic acid	2, 4-D	Potato and rice	0.1-0.5
Organophosphorous	Anilofos	Rice	0.01

SAMPLE PRETREATMENT AND QUANTIFICATION

Conventional Techniques for Sample Preparation

Determination of herbicide residues in various matrices through sample preparation and instrumental analysis is a challenging task as very low MRLs have been imposed by regulatory agencies, and therefore ultra-sensitive analytical methods for sample preparation are of utmost importance. In the determination of herbicide residues, the most important step is the sample pretreatment which is comprised of isolation and concentration of analytes with minimum matrix interferences, thus allowing easy detection and quantification through instrumental analysis [27].

Initially, a bioassay was used to detect the phytotoxicity effect of both active substances and the possible transformation products of the herbicide in soils. In ecotoxicology studies on the residue of different herbicides and their transformation products, the most frequently applied species include plants and their seeds. On the basis of the dose-response curve, sensitive indicator plants for quantitative measure of phytotoxicity of the tested substance are determined. The results of the bioassay are expressed in terms of ED_{10}, ED_{50}, ED_{90} (effective dose) or IC_{50}, IC_{90} (inhibition concentration). It is an economical and facile technique that does not require expensive organic solvents and instruments and is the preferred method for the estimation of phytotoxic herbicide's activity. The European Commission recommends the use of bioassays as an acceptable method for their ability to detect herbicide residues at low levels in soil [28]. It is a suitable screening test useful to exclude the occurrence of low levels of residues of phytotoxic compounds, and the results of these bioassays are used to guarantee non-injury to the succeeding crop in crop rotation [29]. Different herbicides and their corresponding sensitive indicator species are given in Table **3**.

Table 3. Common injury symptoms of herbicides on sensitive species.

Herbicides	Bioassay/Indicator Species	References
2,4-D	Cotton, pigweed, tomato, mustard	[30]
Dicamba	Beans, sorghum, cucumber	[31]
Imazaquin and Imazethapyr	Cotton, Sugar beet, canola, mustard	[32, 33]
Chlorimuron, Thifensulfuron, Nicosulfuron, Flumetsulam, chlorsulfuron	Cucumber, sugarbeet, Oats, lentils, Sugar beet, canola, mustard	[34 - 36]
Sethoxydim, Clethodim	Soyabean, cotton, sunflower	[37, 38]
Trifluralin, fluchloralin, ethalfluralin	Oats, sorghum, rice, cucumber, wheat, barley	[39, 40]
Pendimenthalin, Metolachlor, Acetochlor, Dimethenamid	Algae and aquatic plants, ornamental crops	[41 - 43]
Atrazine, Simazine, Metribuzin	Oats, cucumber, sugarbeet	[44, 45]
Linuron	Maize, sunflower, ryegrass	[46]
Fomesafen, Flumioxazin	Cotton	[47, 48]
Alachlor, Metachlor	Cucumber, ryegrass, crabgrass	[49, 50]
Chlorpropham	Cucumber, oats, ryegrass	[51]
Diuron, Monuron, Isoproturon	Cucumber, oats, barley, millet, sorghum	[52 - 54]
Picloram, Clopyralid	Faba bean, flax, lentils, peas	[55, 56]

In some studies, bioassay has been reported to be more sensitive than instrumental analysis. Paul *et al.* [57] observed that the bioassay technique for quantification of metsulfuron-methyl residues using lentil seeds was more sensitive than HPLC and was a large difference in residues detected by both techniques. This may be due to the strong binding of metsulfuron-methyl to soil over time and growing plants released bound residues from the soil, resulting in improved detection of residues.

However, major drawbacks of bioassay include that it could not quantify residues from food, has variable sensitivity due to variation in plant growth, lacks reproducibility, and the choice of indicator species is determined by herbicide sensitivity to that assay. Therefore, different techniques for herbicide residue, quantification of herbicides, *viz.* liquid-liquid extraction (LLE), solid-liquid extraction (SLE) and soxhlet extraction, have been adopted for extraction from soil and food to increase the concentrations of the analytes (enrichment) to a level above the detection limit for a given analytical technique.

The first step in the pesticide residues analysis from semisolid and solid samples is usually the exhaustive extraction of the target compounds from the matrix in which they are entrapped. The non-selective character of this initial treatment makes the subsequent purification necessary so as to eliminate the interfering matrix. LLE has been used for a long time for the determination of herbicide residues, and good accuracy and precision have been achieved using this method (Table **4**).

These are time-consuming, labour intensive, expensive and multistep processes. Sometimes emulsions are formed during the extraction process, which reduce the extraction efficiency of the analyte. In addition, there is a requirement for a large amount of matrix. Recent regulations concerning the use of organic solvents have made classical methods unacceptable because of large solvent consumption.

Considering this, the latest trends in green analytical chemistry include the miniaturization of LLE techniques [70] which helped to recognize the importance of extraction technology in the generation of quality analytical information [71]. Miniaturization techniques for quantification of herbicide residues involve a reduced volumetric ratio of extractant phase to sample, faster sample preparation, low cost, extremely low or even no solvent consumption, reduced generation of wastes, high extraction efficiencies and improved automation properties.

Table 4. Quantification of herbicides using LLE.

Analyte	Matrix	Specific Conditions	LOQ (µg/mL)	Recovery (%)	References
Atrazine (P)	River water	Methanol[a]/15 min[b]/ Ethyl acetate[c]	0.001	92.0	[58]
Imazosulfuron (P)	Food, water and soil	Ethyl acetate/75 min/ dichloromethane (DCM)	0.015	70-90	[59]
Atrazine and simazine (P); 2-hydroxuterbuthylazine (TP)	Honey	Acetone/60 min/ACN-water (10:90)	0.3-1.5	78-82	[60]
2,4-D, atrazine and alachlor (P)	Water	Acetone/30 min/chlorobenzene	15×10^{-5} - 0.003	82-107	[61]
Five triazine herbicides (P)	Water and soil	Chloroform/5 min/CAN	15.9×10^{-5} -0.0006	82-102	[62]
Flumetsulam	Soyabeans	DCM/30 min/methanol:hexane (4:1)	0.051	72-92	[63]
Butachlor	Water	Water/30 min/DCM	0.0001	81.0-93.0	[64]
Pendimethalin	Water	Water/30 min/DCM	0.0001	76.0-86.9	[65]

(Table 4) cont.....

Analyte	Matrix	Specific Conditions	LOQ (µg/mL)	Recovery (%)	References
Pretilachlor	Paddy water and sediment	Methanol/45 min/DCM	0.03×10^{-2}	76.6-81.0	[66]
Pretilachlor	Water	Water/30 min/DCM	0.0016	81.6-97.6	[67]
Anilofos (P)	Soil and Plants	ACN:water/45 min/ DCM	0.01	> 95	[68]
	Soil, grain and straw	chloroform:acetone (1:1,v/v)/60 min/DCM	0.007	83.1-94.3	[69]

[a]Extracting solvent [b]Extraction time [c]Partitioning solvent.

Miniaturization of LLE Techniques

In miniaturization of LLE *viz.* liquid phase microextraction (LPME), the extraction phase is liquid, and the volume of the extractant phase is restricted to a few microliters. These extraction techniques require a small sample volume, and the entire volume is injected into the analytical system. Different LPME strategies used for extraction of herbicide residues and their transformation product include single drop microextraction (SDME), hollow fibre liquid-phase microextraction (HF-LPME), dispersive liquid-liquid microextraction (DLLME) and salting out liquid-liquid extraction (SALLE) (Table **5**). SDME is a simple, low-cost, fast and virtually solvent-free sample preparation technique and four different types of SDME techniques, *viz.* direct single-drop microextraction (direct-SDME), headspace single-drop microextraction (HS-SDME), liquid-liquid-liquid microextraction (LLLME) and continuous-flow microextraction (CFME) have been used for the extraction of herbicides of a wide range of polarity from different matrices (Table **5**). Factors influencing the extraction procedure like type and volume of extraction solvent, extraction temperature and time, sample stirring rate, pH of sample solution and salting out effect were optimized. Developed methods were simple, inexpensive and recognized as environmentally benign as these are virtually solvent-free, requiring only a fraction of a drop of organic solvent.

In direct SDME, fiber is pushed out of the hollow needle and is immersed into the sample directly, while in the headspace (HS-SDME), the sorbent coating is exposed to the headspace of the sample where the target analyte is present. After the sorption of the analyte in both cases, the fiber is drawn into the needle; the needle is withdrawn from the sample vial and transferred to the injection port of the analytical instrument. A major drawback of direct-SDME is the instability of drop at high stirring rates and temperature, particularly when samples are not cleaned perfectly. In addition, solvents with low boiling points and higher water solubility are not ideal for direct SDME due to high dissolution or evaporation rates. HS-SDME was introduced to remove the limitations of drop instability and limited surface area in direct-SDME.

The drawbacks of dislodgement and instability of droplets in SDME due to stirring speed and complicated matrix have been overcomed by the use of ionic liquids (IL). IL has high thermal stability, good solubility for organic and inorganic compounds, and high viscosity so that they can form large drops and remain stable during extraction resulting in better extraction efficiency and high enrichment factor (Table **5**). Magnetic ionic liquids (MILs) have been used widely in the extraction of triazine herbicides in SDME [72], where a magnetic rod is used to hold a hanging drop of MIL, exposed to the sample solution or the headspace. This method allows the use of large and stable drops for longer extraction times, even under strong conditions of stirring.

LLLME is suitable for ionisable analytes, and is based on the extraction of analytes from the aqueous stirred sample into an organic layer with a density lower than water and simultaneous back-extraction into an aqueous micro drop. In CFME, a glass extraction chamber is used to carry out extraction, and instead of stirring, the sample is circulated at a steady flow rate, and when the extraction chamber is full of samples, a drop is produced at the tip of the microsyringe needle.

Table 5. Quantification of herbicides using different miniaturized LLE techniques.

Analyte	Matrix	Specific Conditions	LOQ (µg/mL)	Recovery (%)	References
Direct SDME					
MCPB, MCPP and triclopyr (P)	Water	n-heptane[a]/25 min[b]/Na$_2$SO$_4$[c]/Thrombin binding aptamer (TBA)[d]	1.8-3.0	84.8-107.9	[73]
Metobromuron, monolinuron, linuron (P); 4-bromoaniline, 4-chloroaniline and 3,4-dichloroaniline (TP)	vegetable	Water/15 min/NaCl/polyacrylate (PA) fiber	0.0005-0.002	> 79	[74]
Atrazine (P); DEA and DIA (TP)	Water	1-octanol/20 min/NaCl/hollow fiber	1×10^{-5}-5×10^{-5}	65.6-96.3	[75]
Ametryn and atrazine (P)	Mango	ACN/15 min/Na$_2$SO$_4$/hollow fiber	0.003-0.004	69-113	[76]
Alachlor, acetochlor, metolachlor, pretilachlor and butachlor (P)	Water	Toluene/15 min/NaCl/ PA fiber	2×10^{-7}-0.0001	80-102	[77]

(Table 5) cont.....

Analyte	Matrix	Specific Conditions	LOQ (μg/mL)	Recovery (%)	References
Diazinon, chlorpyrifos-methyl, chlorpyrifos-ethyl and endosulfan (P) Endosulfan sulphate (TP)	Honey	Acetone/40 min/NaCl/ Polytetrafluoroethylene (PTFE) based fiber	0.0001-0.005	73.5-119	[78]
HS-SDME					
Butanone (P)	Milk powder	Hexadecane[a]/9 min[b]/potassium ferrocyanide and zinc acetate[c]/PTFE silicon based[d]	0.002-0.009	89.4-107	[79]
Chlorpyrifos-ethyl (P)	Rat liver	1-octanol/10 min/anhydrous sodium sulphate/porcelain based fiber	0.001	> 98	[80]
CFME					
Simazine (P)	Water	Ethyl acetate[a]/10 min[b]/ NaCl[c]/ hexachlorobenzene and opentachlorobenzene[e]	0.004	104-106	[81]
DLLME					
Alachlor, pendimethalin and diphenylamine (P)	Wine	Chloroform/40 min/acetonitrile	2.5×10^{-5}-0.0002	85.9-93.9	[82]
2,4-D, MCPB, fluazifop and haloxyfop (P)	Water	Trichloromethane/30 min/1,4-dioxane	0.0002-0.0009	> 80	[83]
Diclofop-methyl and fenoxaprop-P-ethyl	Fruits	1-octanol/45 min/Acetic acid	0.0005-0.0008	92-100	[84]
HF-LPME					
Atrazine (P) DEA, DIA, DDA and ATOH (TP)	Water	di-n-hexylether/5 hrs/NaCl/polypropylene hollow fiber	3×10^{-5}-0.0011	> 90	[85]
Carbendazim, thiabendazole, fuberidazole, triazofop and carbaryl (P) 2-aminobenzimidazole, and 1-naphthol (TP)	Soil and water	1-octanol/30 min/ NaCl/ polypropylene hollow fiber	1.06×10^{-6}-0.0073	85-117	[86]
Flumetsulam (P) and TP	Soil	DCM/30 min/NaCl/PA fiber	0.0005-0.0008	85.2-115	[87]

(Table 5) cont.....

Analyte	Matrix	Specific Conditions	LOQ (µg/mL)	Recovery (%)	References
SALLE					
chlorosulfuron, foramsulfuron, nicosulfuron, oxasulfuron, prosulfuron, triasulfuron and triflusulfuron-methyl (P)	Water and banana juice	ACN[a]/10 min[b]/(NH$_4$)$_2$SO$_4$[c]/methanol[e]	0.0004-0.013	72-115	[88]
Atrazine, secbumetone, terbutryn, ametryne and Aziprotryne (P); DIA, DEA and ATOH (TP)	Water	ACN/15 min/NaCl/DCM	2×10^{-5}-0.0001	62.5-114.6	[89]
Paraquat (P)	Food	ACN/10 min/NaCl/DCM	0.02	80-96	[90]

[a]Extracting solvent [b]Time [c]Salt [d]Fiber [e]Partitioning solvent [f]Dispersive solvent.

HF-LPME is a mode of LPME that uses a porous polypropylene hollow fibre for immobilization of organic solvent in the pores of hollow fibres. The key components of the procedure are the donor phase, which is typically an aqueous sample containing analytes of interest, porous polypropylene hollow fiber for the immobilization of the organic solvent in its pores, organic solvent immobilized in hollow fibre pores, and acceptor phase, which is usually an organic, basic or acidic solution that fills the hollow fibre lumen from inside. The extraction efficiency of the desired analyte depends upon the type of hollow fibre material, pH of acceptor and donor phase, extraction time, extraction temperature, type of organic solvent and addition of salt. Commonly used hollow fibre materials include polysulfone, poly (vinylidene fluoride), polyethylene, polydimethylsiloxne and the solvent used should be non-volatile, immiscible with water, strongly immobilized within the pores of hollow fibre, able to provide high solubility for the target analyte. pH of the donor phase should guarantee analyte neutrality and thus reduce the sample solubility in solution, while that of the acceptor phase should guarantee the ionization of the analyte. The addition of salts like sodium chloride, potassium chloride and sodium sulphate enhances the extraction efficiency due to the salting out effect. HF-LPME technique followed by LC-MS analysis has been used for the extraction of triazine, sulfonamides and benzimidazoles with high extraction efficiencies [91] due to the enhanced mass transfer process under vigorous stirring during the extraction (Table **5**).

In DLLME, a solvent miscible with sample and extractant solvent is added in order to promote cloudy mixture formation. This enhances the transference of analytes and contact surfaces, and avoids problems such as drop dislodgment. The type and volume of extraction and disperser solvent, extraction time, pH and addition of salt are the most important part of this technique. The extraction solvent should be miscible with disperser solvent and has low solubility in water to enable the formation of fine droplets of extraction solvent in the aqueous phase. Solvent with a density greater than that of water and have a tendency to form a cloudy solution with the dispersive sorbent. Trisulfuron, metsulfuron methyl, chlorosulfuron, primisulfuron-methyl, flazasulfuron and chlorimuron ethyl [92], triazine (atrazine), chloroacetamide (alachlor), and phenoxy (2,4-dichlorophenoxyacetic acid) [93] have been extracted using DLLME with high efficiencies (Table 5).

SALLE is a simple pre-treatment technique with proven efficiency for the extraction of herbicides from water [94]. It is based on liquid-liquid extraction using an appropriate amount of salt to achieve the separation between the aqueous phase and partially miscible organic phase, resulting in simultaneous extraction of target analytes into the separated organic phase. SALLE method has been proposed for the determination of sulfonylurea, triazines herbicides in environmental water and banana juice samples [94, 95]. Various parameters affecting the extraction process, such as the type and volume of the organic solvent, type and amount of salt, pH of the sample and vortex time, were optimized. Organic solvents like isopropanol, ethyl acetate, acetonitrile, acetone and acetone:acetonitrile have been commonly used for extraction of herbicides depending on physico-chemical properties of herbicides. The addition of salt decreases the solubility of hydrophilic compounds in the aqueous phase through a salting-out effect and consequently increases the partition of analytes into the organic phase. Different salts, such as sodium chloride, ammonium sulphate and sodium carbonate, have been used in SALLE depending upon their different degree of phase separation. Sodium chloride is widely used in SALLE due to more accurate results in terms of reproducibility and extraction efficiency [96, 97, 98]. This might be due to greater salting-out ability and more solubility of sodium chloride in water as compared to other salts.

Solid Phase Extraction (SPE)

SPE is a commonly used technique due to its rapidity, simplicity and ability to treat huge volumes of samples with maximum recoveries. It has been successfully used to simplify complex matrices, purify compounds, reduce ion suppression in mass spectrometry, concentrate analytes present at low levels and fractionate complex mixtures. It is based on the partitioning of the solute between two phases, a liquid (sample matrix or solvent containing analytes) and a solid (sorbent) phase.

The type of sorbent in SPE is chosen on the basis of the chemical properties of the analyte that is to be separated. Commonly used sorbent material includes normal phase (silica, alumina and cyanopropyl (CN) SPE cartridges), reverse phase cartridges such as C_{18} (Sep-Pak Plus) and C_8, ion exchange cartridges like amino (NH_2), primary secondary amine (PSA), strong cationic SPE sorbent *viz.* propylsulfonic acid (PRS), strong anion exchange (SAX) SPE cartridge *viz.* ion-exchange SAX SPE cartridges and mixed mode or multiple mode cartridges (Table **6**). Mixed-mode and multidimensional SPE combines the advantages of reversed-phase and ion-exchange SPE for enhanced separation, and analyte is retained through a primary and secondary mechanism. Some mixed-mode SPE cartridges include reversed-phase/strong anion-exchange, reversed-phase/weak anion-exchange, reversed-phase/strong cation-exchange and reversed-phase/weak cation-exchange. The mixed-mode SPE cartridge (Oasis MAX) is able to retain and isolate a wide range of herbicides with acidic, neutral and basic characteristics from a single matrix. Selective elution and highly reproducible recovery of ionically bound analytes are attained by manipulating the charge of analyte and the sorbent (Table **6**).

Table 6. Quantification of herbicides using SPE.

Analyte	Matrix	Conditions	LOQ (µg/mL)	Recovery (%)	References
Penoxsulam (P), Sulfonamide, 2-amino TP (TP)	Rice	ACN:methanol:acetic acid[a]/Oasis HLB[b]/40 min[c]	0.005-0.009	> 95	[99]
Propoxycarbazone (P)	Soil and water	ACN/Oasis HLB/30 min	0.001-0.01	74-113	[100]
Atrazine, terbutryne and chlorotoluron (P); DEDIA, DIHA, 2-hydroxyterbutylazine and 3-chloro-4-methylphenylurea (TP)	Surface and ground water	Methanol/Styrene-divinylbenzene/45 min	6×10^{-5} - 0.0003	68-109	[101]
Nine triazines (P); desethyl atrazine, desethyl-desisopropyl atrazine, desethyl2hydroxyatrazine, desethyl terbuthylazine (TP)	Water	Online SPE water/methanol/Oasis HLB/11 min	2.3×10^{-5} - 0.0006	81-99.8	[102]
4-CPA, 2,4-D, dicamba and 2,4-DP(P)	Water and cucumber	ACN/Cotton@UiO-66 based/20 min	0.0003 - 0.001	91.7 - 105.4	[103]
Metaxuron, monouron, chlortoluron, monolinuron and buturon (P)	Tomato and milk	ACN/Py-DMB HCP packed/40 min	0.0002-0.0006	86-111.7	[104]
Atraton, Simetryn, Prometon, Ametryn, Propazin and Prometryn (P)	Corn	n-hexane/MWCNTs/40 min	3.71×10^{-5}	90.30 to 116.24	[105]

(Table 6) cont.....

Analyte	Matrix	Conditions	LOQ (µg/mL)	Recovery (%)	References
Tribenuron-methyl (P)	Water, soil and soyabean	ACN/Oasis MAX/7 min	9.54×10^{-7} - 4.77×10^{-6}	84.6-106.7	[106]
Chlorsulfuron (P)	Water	Online SPE ACN/MIPs/30 min	-	81-110.1	[107]
Mesotrione (P); AMBA and MNBA (TP)	Water and soil	DCM/Oasis SDB/35 min	0.06	90.3	[108]
Pinoxaden (P); Dihydroxy metabolites (TP)	Soil, water	DCM/Oasis HLB/25 min	0.6	98-100	[109]
Triafamone (P); Triafamone-dihydro and oxazolidone-dione (TP)	Wheat	0.1% acetic acid/Silica gel/25 min	0.05	84-94	[110]

[a]Extracting solvent [b]cartridge [c]Extraction time.

Polymer-based sorbents such as styrene-divinylbenzene (SDVB) have also been used in SPE with the advantages of not requiring acid / basic elution modifiers nor presenting pH limitations as these are stable at a wide range of pH (1 to 14). It also has high sample capacity and great flexibility during the development of the method. Biesaga *et al.* [111] reported the extraction of phenoxyalkanoic acid herbicides using C_{18} BondElut, phenyl-silica, LiChrolut SAX and polymeric sorbents and more than 95% recoveries were obtained with polymeric and phenyl-silica sorbents using pure methanol for elution. Simultaneous extraction of 30 sulfonylurea herbicides from water samples was done using LC-MS using N-vinyl-pyrrolidone polymeric cartridge (Oasis HLB) with LOQ ranging from 0.1-5.9 ng/L and 0.4-5.8 ng/l, respectively in tap and leaching water [112].

Recently, the use of new and selective materials such as nanofibers, carbon nanotubes, immobilized receptors or antibodies (IMS), molecularly imprinted polymers (MIP) and restricted access materials (RAM) are used as sorbents which have resulted in increased sensitivity and selectivity of SPE. MIPs are highly cross-linked polymers optimised to isolate a single analyte or class of structurally similar analytes with high selectivity. MIPs contain cavities that can bind target analyte through multiple interaction points (reversed phase with polymer backbone, ion-exchange and hydrogen bonding). Hence, the MIPs binding sites entail a strong interaction with the analyte requiring harsh eluting conditions and producing cleaner extracts. As a result of their extreme selectivity, MIPs reduced the matrix effect and provided low background noise and thus lower LODs. MIPs also offer advantages such as batch sample processing capabilities, small size, and adaptability to robotic technology, low cost, and ready availability from many sources. Mirzajani *et al.* [113] fabricated flexible MIP-SPME fiber using fused silica capillaries as mold and applied it to simply, selectively and accurately determine the presence of thidiazuron on fruit and vegetable samples without any derivatisation steps using an ion mobility spectrometer. The method has high sensitivity, good quantification extraction efficiency, wide linear range than other techniques. The consumption of toxic organic solvents was minimized without affecting method sensitivity and provided good purification with no interference peaks.

Multi-walled carbon nanotubes (MWCNTs) have also been used for the extraction of sulfonylurea (nicosulfuron, thifensulfuron methyl and metsulfuron methyl) and triazine herbicides (atrazine and simazine) in environmental water [114], atrazine and its metabolites *viz.* DIA and DEA in water and soil [115]. Efficient recovery was obtained for herbicides, and the method was competitive for trace level analysis of studied herbicides. Two novel materials based on periodic mesoporous organosilica (PMO) with cationic aminebridged ligands, (styrylmethyl)bis (triethoxysilylpropyl)ammonium chloride (PMO-STPA) and bis(3- triethoxysilyl propyl)amine (PMO-TEPA) with reverse-phase/strong anionic exchange mixed-mode or strong anionic exchange retention mechanism, respectively have been used for enantiomeric separation and extraction of six phenoxy herbicides (fenoprop, mecoprop, dichlorprop, 2-(4-chlorophenoxy) propionic acid (4-CPPA), 2-(3-chlorophenoxy) propionic acid (3-CPPA), 2-phenoxypropionic acid (2-PPA)) from water [116].

In addition to sorbent, the elution solvent plays an important role in increasing the clean-up efficiency by disrupting the interactions between the target analyte and the adsorbent. Organic solvent has been used either single or mixture for extraction and elution of the herbicide residues. The choice of appropriate solvent depends on the molecular characteristics (non-ionic/ionic) of herbicides. Solvents *viz.* ACN, ethyl acetate, methanol, DCM, acetone, hexane, petroleum ether, diethyl ether, acetic acid, toluene and cyclohexane have been used.

The technique can be performed in both offline and online modes. Off-line methods provide higher flexibility and do not suffer from the limitation that final conditions of elution from SPE cartridges need to be compatible with the detection system [117]. In automation or online SPE, cartridges and samples are processed in a completely enclosed system. Direct elution is performed into the detection instrument *via* transfer line, and this eliminates erroneous steps like evaporation and reconstitution, making online SPE more efficient and fully automated. On-line SPE reduces chemical waste due to the use of a minimal amount of extraction solvents with good precision and recovery.

SPE offers a series of advantages such as less consumption of the solvent required, improved precision and accuracy with no cross-contamination, less exposure to toxic agents and higher recoveries with minimal sample transfer, shorter analysis time and is applicable to a wide range of herbicides having different physicochemical properties. Although better separation and recovery from complex matrices can be obtained through SPE, due to its perceived difficulty in mastering, its usage has become a tedious process. It requires expensive SPE cartridges, which cannot be regenerated and reused for the extraction of analyte. Also, repeated extraction is required for the extraction of strongly adsorbed analytes which may plug the SPE membrane, decreasing the flow rate and hence resulting in a reduction of extraction efficiency. Besides many stages involved in carrying out this technique, it may facilitate contamination of the sample. SPEs have the disadvantages of being unproven for many herbicides, being unable to handle large sample sizes, and generally being ineffective for extracting water-soluble herbicides and metabolites.

Miniaturization Techniques of SPE

To overcome these limitations, novel microextraction techniques were introduced, which require less time and labor. These micro-extraction techniques allow the integration of activities (*e.g.*, sampling, extraction and enrichment) of analytes to a level above LOD and isolation of analytes from the sample matrix [118]. Currently, different miniaturized sample preparation and concentration techniques such as solid phase microextraction (SPME), stir bar sorptive extraction (SBSE), dispersive micro-SPE (DMSPE) and magnetic SPE (MSPE) have been exploited for the extraction of herbicides from a variety of matrices (Table **7**).

SPME is a non-exhaustive sample preparation technique in which a small amount of extraction phase in the range of microlitres is used. It is a single-step extraction procedure involving analyte isolation and enrichment. Its advantage over SPE includes the small sample size, reduced sample preparation time, high sensitivity, low cost, eliminated the use of expensive and toxic organic solvents and improved detection limits. In this, thin fibers coated with an appropriate sorption material are placed in contact with the sample matrix by direct immersion or headspace, and after the extraction step, the SPME fiber is transferred to the injection port of the instrument. Commonly used extraction phases include polydimethylsiloxane and its derivatives, graphene, porous carbon, molecularly imprinted polymers (MIPs), metal-organic framework and polymeric IL (Table **7**). Peak broadening is smaller in SPME due to the complete desorption of analytes before injection, and precision and accuracy depend on the quality of the fibers. Fiber degradation can occur during repeated use, and the system is relatively fragile and easily broken while handling. SPME is relatively expensive and has the possibility of transferring between analyses.

The use of an internally coated capillary or needle as an alternative to coated fibre is the base of the in-tube SPME technique. In-tube SPME uses open-tubular capillary columns for retention of the analyte. This technique was developed primarily to overcome some problems related to the use of conventional fiber SPME, such as fragility, low sorption capacity and bleeding of thick film coatings of fiber [119]. The technique has a principle similar to fiber SPME, but in this extraction of analyte is performed on the inner surface of the capillary column, while in fiber, SPME extraction is performed on the outer surface of the fiber. Thus, plunging of the capillary column and flow line during extraction is to be prevented by pre-filtration of samples.

Chlorinated phenoxy acid and triazine herbicides in water samples have been extracted using in-tube SPME, in which herbicides were automatically extracted into a DB-WAX capillary [120]. Recoveries for chlorinated phenoxy acid and triazine herbicides in water samples were above 95%. The developed method can continuously extract herbicides from aqueous samples providing a simple, rapid, selective, sensitive, low-cost method for herbicide analyses and can be directly applied to various water samples without any pretreatment.

The principle of SBSE is the same as SPME, where solutes are extracted into a polymer coating on a magnetic stirring rod (PDMS). A magnetic rod is encapsulated in a glass jacket on which PDMS coating is placed. Extraction is controlled by the partitioning coefficient of solutes between the sample matrix and polymer coating. During extraction, the solute migrates from the sample into PDMS coating, and the uptake rate is controlled by a diffusion constant, volume and size of stir bar coating, stirring condition and sample volume, pH, addition of salt, temperature and addition of organic solvent and subtle interplay between these. SBSE method using polyethylene bar impregnated with activated carbon has been used for the quantification of triazine [121], phenylurea and ditroaniline herbicides [122] in environmental water samples, and this method could be considered as an alternative to more conventional extraction procedure such as SPE for rapid screening of water contamination.

In contrast to SPME, where desorption is performed in the inlet of gas chromatography, SBSE is used in combination with thermal desorption (TD) system for optimum desorption, re-concentration and GC analysis. The process of desorption is slower than that of SPME fibre. In-situ and derivatisation of the source after extraction has been performed to increase recoveries and to improve chromatographic performance and delectability. Kawaguchi et al [123] reported an SBSE method with *in situ* derivatization with acetic anhydride and GC-MS-TD for the determination of trace amounts of chlorophenols *viz.* 2,4-DCP, 2,4,6-TrCP, 2,3,4,6-TeCP and PCP in tap and river water. The proposed method involves simplicity of extraction, freedom from use of organic solvents, high sensitivity with recoveries >95%, and precision was < 10%.

DMSPE involves the addition of sorbent to the water sample to form a dispersion. The sorbent used has been derivatized to produce a bound organic phase (*e.g.*, octadecyl, MIP, *etc.*) on its surface. The contact between the analytes and the support is higher than in traditional SPE, increasing the equilibrium rate and providing higher extraction yields (Table 7). This process not only absorbs less solvent and labour, but also prevents cartridge channelling, clogging and analyte loss due to breakthroughs. So far, C_{18}, PSA, GCB, florisil and a combination of these have been used in d-SPE sorbent to provide good clean-up with high recoveries [124].

MSPE is based on the extraction of different compounds from samples using solid with magnetic properties. In MSPE, a magnetic adsorbent is added to a solution or suspension containing the target analyte. The analyte is adsorbed onto the magnetic adsorbent, and then the adsorbent with the adsorbed analyte is recovered from the suspension using an appropriate magnetic separator. The analyte is consequently eluted from the recovered adsorbent and analysed. The mechanism of separation depends on the form of interactions between analyte molecules and the surface functional group immobilised on the induced dipole of the magnetic core. Magnetic particle type and amount, particle size, herbicide interaction period, elution time, desorption solvent and its amount significantly affected the extraction efficiency.

IL functionalised silica as a sorbent followed by LC-MS has been used for extraction and preconcentration of sulfonylurea herbicides. This is a time-consuming method where sorbent is used in cartridge mode, often resulting in tedious column packing, high backpressure and low flow rate. These limitations can be eliminated by providing magnetism to IL-functionalised silica followed by magnetic separation. The magnetic nanoparticles based on N-methylimidazolium IL and Fe_3O_4 have been used for the extraction of sulfonylureas herbicides from polluted water samples [125]. The method allows the determination of herbicides and LOD ranging from 1.13 and 2.95 ng mL^{-1}.

MIL-modified multi-walled carbon nanotubes (MIL-MWCNTs) and imidazolium-modified carbon nanotubes have been used as the sorbent to simultaneously extract aryloxyphenoxy-propionate herbicides (AOPPs) and their polar acid metabolites due to the excellent π-π electron donor-acceptor interactions and anion exchange ability (Table 7).

Matrix Solid-phase Separation

MSPD is an efficient, simple and flexible technique for the separation of herbicides from various plant, animal, biological and soil matrices. In MSPD, extraction and cleanup are combined in a single step, thus making the procedure easy and faster with less consumption of solvent and sample loss. MSPD is based on principles of both chemistry and physics, involving the application of forces to the sample by mechanical blending in order to completely disrupt the sample [142]. This disruption of the sample is induced through shearing and grinding forces. The sample is dispersed over bonded phase surface by hydrophilic and hydrophobic interactions, and this blend is transferred into a glass column. Then, the column was eluted with a suitable solvent for the extraction and isolation of analytes. In addition to dispersing the sample matrix and facilitating the extraction process, the materials which can improve the extraction efficiency and simultaneously clean up the final extract involve lipophilic sorbent material such as non- modified solid support like silicates or other organic (graphatic fibre), inorganic florisil, alumina, C_{18} and C_8 bonded silica and non-retentive supporting materials (*e.g.*, diatomaceous earth and sand) [143] (Table **8**). Co sorbent is added to the column to be filled with blend in some cases to assist the clean-up and fractionate the components of the analyte. The recent trend involves the use of MIPs and MWCNTs after suitable functionalisation [144]. Recoveries obtained using MIPs are better than conventional sorbent materials, *viz.* C_{18}, silica, florisil and sand. MWCNTs are a kind of carbon-based nanomaterial having an excellent adsorption ability because of their hydrophobic surface and structure of internal tube cavity, and extracts obtained using MWCNTs were cleaner than those obtained using C_{18} and diatomite [145].

Table 7. Quantification of herbicides using miniaturised SPE techniques.

Analyte	Matrix	Specific Conditions	LOQ	Recovery (%)	References
SBSE					
Simazine, atrazine, prometon, ametryn, prometryn(P)	Maize	Acetone/Oasis HLB/50 min	4.93-11.9	91-115	[126]
Simazine, atrazine, metolachlor, acetochlor, diuron and chlorpyrifos-ethyl (P); diuron- 3,4-dichloroaniline and 1-(3,4-dichlorophenyl)-3-methyl urea (TP)	Surface water	Methanol:acetonitrile (ACN)/Oasis HLB/40 min	0.2-20	> 80	[127]

(Table 7) cont.....

Analyte	Matrix	Specific Conditions	LOQ	Recovery (%)	References
Simazine, atrazine, terbumeton, terbuthylazine, metribuzin, simetrine, isometioazine and metamiron (P); DIA, DEA (TP)	Water	Methanol/ Oasis HLB /60 min	0.0007-0.011	94.4-106	[128]
MSPE					
Terbuthylazine, secbumeton, terbumeton, atraton, atrazine, prometon and trietazine (P)	Rice	n-hexane/Fe_3O_4 nanoparticles/ 25 min	5×10^{-5} - 0.0002	90.5-105.7	[129]
Nicosulfuron, chlorsulfuron, pyrazosulfuron-ethyl, metsulfuron- methyl and chlorimuron-ethyl (P)	Water	ACN-water/MWCNTs/8 min	3×10^{-5} - 0.001	76.7-106.9	[130]
Sulfosulfuron, bensulfuron-methyl, pyrazosulfuron-ethyl and halosulfuron-methyl (P)	Water	ACN/Fe_3O_4/PDA/15 min	29.4×10^{-6} - 3.3×10^{-5}	61.3-108.6	[131]
Prosulfuron, chlorimuron-ethyl and triflusulfuron-methyl	Water	Methanol/Fe_3O_4 dioctadecyl dimethyl ammonium chloride silica/30 min	24.9×10^{-5} - 0.0003	82.9-106.5	[132]
DMSPE					
Fenuron, simeton, simazine, atraton, chlortoluron, secbumeton and terbumeton (P)	Vegetable oil	n-hexane/MIL-101/15 min	0.0015-0.003	96.2-102	[133]
30 triazine herbicides	Drinking water	Methanol/PCX/15 min	0.6-60	70.5-112.1	[134]
Ametryn, atrazine, prometryn, simazine and terbutryn (P)	Honey	1% formic acid in ACN:water/PCX/20 min	5×10^{-5} - 0.0005	89.8-116.2	[135]
Haloxyfop acid, quizalofop, cyhalofop acid and dicolofop acid (P); Haloxyfop-methyl, quizalofop-ethyl, cyhalofop-butyl and dicolofop-methyl (TP)	Water	Methanol/MWCNTs/ 25 min	0.009-0.043	66.1-89.6	[136]
SPME					
Acetochlor, alachlor, metolachlor, butachlor and pretilachlor (P)	Water	ACN/Graphite/45 min	18×10^{-5} - 0.0003	81.8-112	[137]

(Table 7) cont.....

Analyte	Matrix	Specific Conditions	LOQ	Recovery (%)	References
Simazine, atrazine, prometon, ametryn, propazine and prometryn (P)	Water	ACN / CNTs/35 min	6×10^{-5}- 0.0003	85-106	[138]
Atrazine, simazine, acetochlor, alachlor and metolachlor (P)	drainage water	Methanol/polydimethylsiloxane/45 min	0.0006-0.0015	> 95	[139]
Atrazine, prometon, ametryn and prometryn (P)	Water	Acetone/grapheme/30 min	15×10^{-5} - 0.0006	88-94	[140]
Thidiazuron (P)	Fruits	Methanol/MIP/40 min	1×10^{-5}	90-130	[141]

[a]Extracting solvent [b]Cartridge [c]Time.

Several factors *viz.* solvent type, sorbent type, average particle size diameter and sorbent:sample weight ratio, have been evaluated for their effect on MSPD extractions. Very small particle sizes (3 to 10 μm) lead to high elution times, and vacuum or pressure is required to obtain adequate flow. The ratio of sample to solid support material most often ranges from 1 to 4, and it varies from application to application. Choice of eluting solvent and sequence and design of an elution profile is optimised so as to elute the target analyte efficiently with a high degree of specificity while removing the target analytes. The selection of solvent for elution is connected with the nature of solid material, and organic solvent and their mixtures are used. There is variation in elution volume for each application, and is optimised to reduce the consumption of solvent and co-elution of matrix components.

Table 8. Quantification of herbicides using MSPD.

Analyte	Matrix	Specific Conditions	LOQ (μg/mL)	Recovery (%)	References
Pretilachlor (P)	Soil and rice	Ethyl acetate/Florisil	0.024	88.7-93.3	[146]
Bispyribac sodium (P) and TPs	Soil	Methanol/Silica	0.006	82.7-105.1	[147]
Mesotrione and atrazine (P); Hydroxyatrazine, deisopropylatrazine, desethyldesisopropylatrazine and deethylatrazine (TP)	Plants	Methanol:0.1 M HCl/ Silica	0.001-0.003	67-95	[148]
Desmetryn, secbumeton, prometon, prometryn and terbutryn (P)	Peanut	ACN/diatomite	0.0016-0.0057	80.4-120	[149]

(Table 8) cont.....

Analyte	Matrix	Specific Conditions	LOQ (µg/mL)	Recovery (%)	References
3-aminopropyltriethoxysilane (APTES), chlorimuron-ethyl, nicosulfuron, thifensulfuron-methyl and chlorsulfuron (P)	Grain	ACN/mesoporous molecularly imprinted polymers (M-MIPs)	20.7×10^{-5}-0.0003	93.8-108.6	[150]
Mesotrione and atrazine (P) Hydroxyatrazine, desethyldesisopropylatrazine and deethylatrazine (TP)	Plants	Methanol:HCl/Silica gel	0.001-0.0031	67-95	[151]
Cyhalofop-butyl	Rice grains	ACN/Neutral alumina	0.007	100.1	[152]

[a]Eluting solvent [b]Sorbent.

Compared with other methods, MSPD eliminates the steps of repeated centrifugation, filtration and cleanup, which dramatically reduce solvent consumption and time. Even though MSPD and SPE have many characteristics in common, the mechanisms involved in both are quite different. MSPD involves complete sample disruption and dispersal onto small size particles, providing more surface area for extraction of sample, while in SPE, absorption of sample is on top of the column. The physico-chemical interactions of components are greater in MSPD and different from SPE. Retention properties in MSPD are a mix of adsorption, partition and paired ion/ paired chromatography.

Ultrasonic Assisted Extraction (UAE)

UAE is a versatile, novel technique applied to extract target molecules from various matrices (soil, food and environmental water samples). Different UAE parameters, such as particle size of the sample, solvent type, sonication temperature and time and solid-to-liquid ratio, are optimized to maximize the extraction efficiency of herbicides and their transformation products (Table 9). Choice of an appropriate solvent for extraction of herbicides is the crucial step in UAE procedure. The application of different extraction cycles improves extraction efficiency in some cases. Temperature affects the extraction efficiency as it influences the solubility of analytes in the medium and cavitation phenomenon, affecting the mass transfer. An increase in temperature enhances the permeation of solvent into the matrix by reducing solvent viscosity, but the co-elution of interference compounds is also enhanced, and target analytes may be degraded. After extraction, an effective clean-up step is performed, especially while analyzing the complex matrices to have good sensitivity and accuracy in the analytical determination of target compounds. When various compounds with different physicochemical characteristics have to be determined, the cleanup of the extract is probably the most challenging step of the analysis. The procedure for clean-up usually prefers solid phase extraction (SPE) using a wide variety of sorbents in bulk or pre-packed in cartridges, columns and pipette tips.

UAE can be performed in tubes, using a bath or probe, followed by centrifugation or the use of columns followed by vacuum filtration for the extraction of analytes. When UAE is carried out in probes, it is called focused ultrasound extraction (FUSE). The power of this extraction technique is 100 times higher than the conventionally used sonicator, and extraction time is significantly shortened. However, the use of probes in extraction may cause cross-contamination and erosion of the probe resulting in a loss in efficiency. Compounds that are volatile may be lost using probe due to more heating of medium and degassing effect. However, higher ultrasonic energy from the horn can sometimes be used to obtain good extraction recoveries.

New solvents *viz.* surfactants, supramolecular solvents and ionic liquids have been used in extraction. In ultrasound-assisted surfactant enhanced emulsification microextraction (UASEME), surfactants increase the contact surface between two immiscible phases, favoring mass transfer of analytes into the organic phase and resulting in improved extraction efficiencies, accelerating small solvent droplet formation, thus reducing the time of extraction. Combining the benefits of microextraction and ultrasound has made it possible to establish an efficient preconcentration technique for determining analytes at low levels. UASEME has been used for extraction of chlorinated phenoxyacetic acids, triazines and sulfonylurea herbicides in the soil, water and fruit samples using chlorobenzene, dichlorobenzene, chloroform, 1-octanol as extracting solvent.

Table 9. Quantification of herbicides using UAE.

Analyte	Matrix	Specific Conditions	LOQ (µg/mL)	Recovery (%)	References
UAE					
Penoxsulam (P); Amino-TP, 5-OH-XDE-638, BSTCA, BSTCA-methyl and BST (TP)	Soil and water	ACN[a]/3 min[b]/ 40°C[c]	0.01	80.1-98.4	[153]
Pendimethalin (P)	Soil and rice grain	Acetone/3 min/30°C	0.003	80.3-101.3	[154]
Imazamox (P); Hydroxyl and glucose metabolites (TP)	Leaves	Methanol: water/3 min/40°C	0.0002	98.7-111.4	[155]
Pinoxaden (P)	Wheat and vegetables	ACN/5 min/30 °C	0.01	90-97	[156]

(Table 9) cont.....

Analyte	Matrix	Specific Conditions	LOQ (µg/mL)	Recovery (%)	References
Simazine, atrazine, isoproturon, linuron, diuron, ametryn and prometryn (P)	Water	ACN/5 min/40°C	15×10^{-5} - 0.0002	66.7-102.3	[157]
UASEME					
Metsulfuron methyl, chlorosulfuron and bensulfuron methyl (P)	Water and soil	1-octanol[a]/1 min[b]/40°C[c]/Aliquot 336[d]	0.0015	103-153	[158]
Metsulfuron methyl, rimsulfuron, tribenuron methyl and primisulfuron methyl (P)	Fruit	Choroform/3 min/30°C/ Tween 80	6×10^{-5} - 0.006	79-105	[159]
IL-UAEME					
Metsulfuron methyl, bensulfuron methyl and pyrazosulfuron ethyl	Soil	[C$_6$MIM][BF$_4$]/ 5 min/40°C	0.027 - 0.039	81.1-100.1	[160]

[a]Extracting solvent [b]Ultrasonication time [c]Ultrasonication temperature [d]surfactant or emulsifier.

In ionic liquid-based UAEME (IL-UAEME), hydrophobic IL is emulsified in an aqueous sample by ultrasound radiation and centrifuged to separate both liquid phases. The application of ultrasound energy acts as a dispersant tool to increase the extraction ability of ILs. This method has been applied for the determination of sulfonylurea herbicides (metsulfuron-methyl, bensulfuron methyl and pyrazosulfuron ethyl) from soil samples using HPLC [161]. Ionic liquid [C$_6$MIM][BF$_4$] was added to the dried soil powder to form a suspension. A water bath at ambient temperature was used to ultrasonicate an analyte with high efficiency and sensitivity. IL- UAEME offers many advantages, such as no use of volatile organic solvent, less time, and low sample and solvent consumption as compared to regular ultrasonication and soxhlet extraction methods.

Supercritical Fluid Extraction (SFE)

SFE is advantageous over the standard LLE as it is carried out with nonpolluting fluids such as CO$_2$. SFE technique offers many advantages such as low consumption of solvent, faster analysis time and environment-friendly nature. Carbon dioxide (CO$_2$), water, methanol, ethylene, propane, *etc.*, are the various types of supercritical fluids. Amongst these, CO$_2$ is the best supercritical fluid used in SFE due to its low toxicity, non-flammability, low cost, high purity, low viscosity and high diffusivity, allowing the efficient wetting and penetration into the solid matrix and ambient critical temperature (31°C) and pressure (72.8 atm) permit extractions of thermally labile non-polar compounds. Other fluids, such as ethylene and methane, having low critical temperatures than CO$_2$, do not dissolve a moderately polar solute to the same extent as CO_2. Similarly, fluids exhibiting high critical temperatures (such as propane, methanol and water) solubilize polar moieties at high concentrations in a fluid phase.

Co-solvents (such as methanol, acetone and ethanol), also known as modifiers or entrainers are added to the supercritical fluid to increase the solubility of the analytes (non- volatile components) in the extracting medium, to improve the separation of the components after extraction, increase in the desorption of the analytes from highly adsorptive sample matrices and to enhance the extraction of the targeted analytes by reducing the co-extraction of the non-targeted analytes (Table **10**). CO_2 is an excellent solvent for non-polar compounds, but it is not good for polar ones. Entrainers are required to increase the polarity of the primary solvent and the extraction efficiency of polar compounds.

The extraction efficiency depends on the solubility of analytes in supercritical fluid and in water. SFE offers a reduction in analysis time as no clean-up of extracts is required. Other than advantages, there are many disadvantages of SFE. The main disadvantage includes the non-polar nature of CO_2, allowing the extraction of mainly non- polar or low polar compounds. Although adding modifiers increases the range of substances to be extracted, adding modifiers may also interfere with the collection step and deactivates the sorbent. This technique has limited application due to the instrumentation required. Evaporation of the solvent at the end of extraction for acquiring a high pre-concentration factor is the critical step in off-line SPE. This is a time-consuming procedure that can cause contamination to the environment and loss/degradation of collected analytes.

Table 10. Quantification of herbicides using SFE.

Analyte	Matrix	Specific Conditions	LOQ (µg/mL)	Recovery (%)	References
Metsulfuron methyl, bensulfuron methyl and chlorsulfuron	Soil	ACN^a/carbon dioxide $(CO_2)^b$/25 min^c/30°C^d	0.0006	> 85	[162]
Atrazine, alachlor, metolachlor, pendimethalin and endosulfan (P); desethylatrazine (TP)	Soil	Ethyl Acetate/ CO_2/ 15 min/90°C	0.5-7.6	80.4-106.5	[163]
Proppham, methiocarb, propoxur and chlorpropham	Soil	Methanol/ CO_2/30 min/60°C	0.01	66-98	[164]
Diclofop-methyl, linuron, metolachlor and quizalofop-ethyl	Honey	Acetone/ CO_2/20 min/60°C	0.018-0.03	87-92	[165]
Pyrazosulfuron-ethyl	Soil	Methanol/ CO_2/60 min/40°C	0.01	> 90	[166]

(Table 10) cont.....

Analyte	Matrix	Specific Conditions	LOQ (µg/mL)	Recovery (%)	References
Trifluralin	Water	Methanol/ CO_2/30 min/40°C	0.5-7.6	80.7	[167]
Atrazine	Soil	Methanol/ CO_2/70 min/30°C	0.01	95	[168]
MCPP, MCPA, 2.4-DP, 2.4-D, 2.4.5-T	Sediments	Online SFE Acetone/ CO_2/25°C	0.003	73-82	[169]
s-triazine herbicides	Onion	Online SFE *n*-heptane/ CO_2/ 53 °C	0.006	80-103%	[170]

[a]Extracting solvent [b]Fluid [c]Extraction time [d]Temperature.

Berglof *et al.* [171] reported the determination of sulfonylurea herbicides (metsulfuron methyl, sulfometuron methyl and nicosulfuron) from soil using methanol as a modifier to enhance the polarity of the targeted analytes. Good recoveries were obtained for metsulfuron-methyl and sulfometuron methyl (75-89%) at 0.4 and 4 $\mu g\ g^{-1}$ fortification levels, whereas only 1 to 4% recovery was obtained for nicosulfuron due to the chemical structure of nicosulfuron as pyridine group in the structure interacts with the surface of the soil and thus, results in the inability to extract the herbicide by SFE method. SFE followed by nanostructured supramolecular solvent microextraction has been used for the extraction of sulfonylurea herbicides *viz.* chlorsulfuron, bensulfuron methyl and metsulfuron-methyl from soil samples and recoveries were found to be higher [172].

Microwave-assisted Extraction (MAE)

It is based on the absorption of microwave energy by polar molecules, which increases the penetration of solvent into the matrix. MAE has been applied for the extraction of organic compounds from various matrices. It is rapid, saves solvent and is efficient in terms of energy use. This method allows the acceleration of energy transfer, facilitates analyte solvation and promotes the disruption of weak hydrogen bonds. Effects of microwave energy depend on various factors, *viz.* nature of solvent, matrix, type of target analyte and sample and solvent dielectric constants.

Choice of solvent is dependent not only on its ability to adsorb microwaves, disrupt interactions with matrix and solubilise the analyte but also on its selectivity towards the analyte of interest, with the exception of other non-labile compounds. Extraction efficiency generally increases with an increase in temperature. However, sometimes extraction efficiency decreases with an increase in temperature due to decomposition of analyte and reaction between analyte and matrix. The dielectric constants of the system are crucial to MAE, since higher dielectric constants promote an increase in the amount of energy absorbed. MAE has been efficiently used for the extraction of alachlor and metolachlor, imidazolinones, triazines herbicides and their metabolites from soil and crop commodities [173 - 176]. It has become an important tool for green analytical chemistry since it significantly reduces solvent volume and extraction times (Table **11**).

Table 11. Quantification of herbicides using MAE.

Analyte	Matrix	Specific Conditions	LOQ (µg/mL)	Recovery (%)	References
Atrazine (P)	Soil and water	ACN: 0.1 HCl[a]/35 min[b]/30°C[c]	6×10^{-5}-0.0009	85-92	[177]
Ametryn, atrazine, desmetryn, prometryn, propazine and simazine (P)	Fruit juices	Methanol/90 min/45°C	0.001-0.02	93-99	[178]
Desmetryn, terbumeton, propazine, terbuthylazine, dimethametryn and dipropetryn (P)	Vegetable	ACN /4 min/ 35°C/quartz as dispersant	0.0015-0.006	76.8-106.9	[179]
Fenuron, monuron, atraton, chlorotoluron, atrazine, ametryn and terbuthylazine (P)	Soyabean	Hexane/4 min/30°C	0.0052-0.0062	91.4-107.6	[180]
Simazine, atrazine, desmetryn, propazine and ametryn (P)	Cereals	Methanol/10 min/ 40°C	0.0035-0.0048	80-102	[181]
Simazine, atrazine, propazine and prometryne (P)	Sheep liver	Methanol/6 min/ 70°C	0.16-0.094	90.3-102	[182]
Atrazine, cyanazine, metribuzine, simazine and deethylatrazine, deisopropylatrazine, acetochlor, alachlor, and metolachlor	Soil	ACN/10 min/40°C	10	> 80	[183]

[a]Extracting solvent [b]Time [c]Temperature.

QuEChERS

QuEChERS (quick, easy, cheap, effective, rugged and safe) is a green, user-friendly extraction and clean-up technique that involves ACN salting out extraction of samples in an aqueous environment followed by d-SPE. The method has been successfully adapted to a number of applications, including the analysis of biological fluids [184, 185], fruits, vegetables and cereals and environmental matrices such as water, sediments, and soil (Table **12**).

In the QuEChERS method, complete homogenization of the target sample is a very important first step. After the target analytes in the homogenized sample are extracted and partitioned using a solvent, they are purified using d-SPE to remove sources of potentially interfering compounds. The amount and combination of solvent and salt are the key parameters that are optimised to increase the extraction efficiency of the target analyte from the complex matrix, reduce the matrix effect and improve the analytical performance of the method. In the original method, extraction was performed under unbuffered conditions [186 - 188]. However, this method is limited in the case of compounds that can degrade at low or high pH. This limitation has been overcome by the use of citrate, acetate, or phosphate buffer and enhanced the extraction efficiency. ACN is commonly used as a solvent in QuEChERS as it extracts the broadest range of organic compounds with co-extraction of lipophilic matrix co-extracts. Additionally, the use of ACN is recommended as, upon the addition of salts, it separates more easily from water than other solvents like acetone and ethyl acetate. Due to the high polarity of ACN, medium to high polar herbicides have better solubility, and hence higher recoveries are achieved when it is used as a solvent.

For separation of phases and salting out effect, various salts such as magnesium sulphate ($MgSO_4$), sodium sulphate (Na_2SO_4), ammonium acetate, sodium chloride (NaCl), sodium acetate and a combination of these have been used to induce ACN/water phase separation during extraction of an analyte. The combination of $MgSO_4$ and NaCl is widely used in QuEChERS as $MgSO_4$ allows the best salting out of ACN, whereas NaCl controls the polarity of the extraction solvent, thus increasing the selectivity of the process. Most favourable extraction was obtained when $MgSO_4$ and NaCl were used in the 4:1 ratio. The probability of $MgSO_4$ to induce chelation has resulted in the use of Na_2SO_4 and NaCl for efficient extraction of pinoxaden and 2, 4 D herbicides [189, 190]. Salts like Na_2SO_4 and $MgSO_4$ tend to deposit solids on the surface of inlet lines of GC-MS source and within the analyser, leading to a loss in instrument performance and linear displacement. These drawbacks have been overcome by the use of salt like NH_4Cl and ammonium formate having high volatility and gets easily evaporated or decomposed in ionization chamber. Concerning the second step in QuEChERS procedure (d-SPE), different sorbents like PSA, octadecyl silica C_{18}, graphitized carbon black (GCB) have been evaluated. PSA is efficient for removal of fatty acids, sugars organic acid, lipids and some pigments while C_{18} effectively removes high lipid content and GCB removes co-extracted pigments namely carotenoids, chlorophyll and other pigmented matrices.

Table 12. Quantification of herbicides using QuEChERS.

Analyte	Matrix	Specific Conditions	LOQ (µg/mL)	Recovery (%)	References
Florasulam and tribenuron-methyl	Soil, plant, grain and straw	1% acetic acid in ACN (v/v)[a]/50 mg PSA, 5 mg GCB[b]/ 20 min[c]	0.005-0.01	72.8-99.2	[191]
Tembotrione	Corn, corn oil and animal foods	ACN/50 mg of C18 and 25 mg GCB/30 min	2	73.7-110.4	[192]
Cloransulam-methyl (P) and TPs	Soyabean and soil	ACN /30 mg PSA+50 mg anhydrous MgSO$_4$/30 min	0.01	87-97	[193]
Clomazone	Soyabeans	ACN with 1% formic acid/C18 + NaCl +MgSO$_4$/25 min	0.005	74.5-110.5	[194]
Penoxsulam (P), 2-amino-TP, 5-OH-XDE-638, BSTCA, BSTCA-methyl and BST (TP)	Soil	ACN/PSA/30 min	3×10^{-6}	87-94.6	[195]
Azimsulfuron, bensulfuron methyl and mesotrione (P) 4-methylsulfonyl-2-nitrobenzoic acid (TP)	Brown rice and rice straw	ACN/MgSO$_4$ NaCl and Na citrate/40 min	0.001-0.05	79-114.5	[196]
Tembotrione (P) 2-(2-chloro-4-mesyl-3-(2,2,2 trifluoroethoxy)methyl)benzoic acid (TP)	Soil	DCM/PSA/20 min	3×10^{-6}	85.6-96.6	[197]

[a]Extracting solvent [b]Sorbents [c]Time.

Hybrid Techniques

The modification of original techniques and combination with other extraction methodologies are done in order to simplify the process and improve extraction efficiencies. These extraction techniques involve reduced solvent and sorbent consumption and help to achieve a more quick, economical and environment-friendly method (Table **13**).

In ultrasonic-assisted MSPD (UA-MSPD), extraction efficiency increased by using an MSPD column in the ultrasonic bath or sono reactor after loading of extraction solvent into the column. Ramos *et al.* [198] reported the UA-MSPD method for triazine herbicides in fruits using reversed-phase octasilyl-derivatised silica (C_8) as dispersant and ethyl acetate as extraction solvent. Low detection limits (1-42 g kg^{-1}) and recoveries above 81% ensure proper determination of herbicide residues at maximum allowed residue levels set in current legislations. Compared with classic MSPD, the proposed method improves the general extraction efficiency, decreases the RSDs and allows complete sample treatment within a few minutes.

Vortex-assisted MSPD (VA-MSPD) is a technique in which the vortex has replaced the column elution step. This technique reduces the consumption of solvent and time of analysis. VA-MSPD has been used in the quantification of herbicides in various matrices (Table **13**). Caldas *et al.* [199] extracted atrazine and clomazone in the fish liver with acetonitrile as eluting solvent and C_{18} as sorbent. In this, the homogenized mixture of sorbent and sample was added to a centrifuge tube, extraction solvent was added, and the tube was vortexed, followed by centrifugation. In comparison to UAE, SPE and MAE, higher recoveries were obtained.

Another modification is magnetically-assisted MSPD (MA-MSPD) and magnetic liquid-solid extraction (MLSE) [200]. For MLSE, analytes in the solid samples are extracted into liquid solution prior to the introduction of magnetic sorbent, while in MA-MSPD, the blending of the solid sample with magnetic nanoparticles was done in order to obtain the homogenous mixture. Both methods have replaced column packing and elution in MSPD with simple magnetic isolation and have not only simplified the extraction process but also reduced extraction time. The magnetic nanoparticles are reusable, thereby providing a solution for the reusability of sorbent in MSPD.

In ultrasonic assisted-dispersive liquid-liquid microextraction (UA-DLLME), extracting and dispersive solvent are mixed before their rapid injection in an aqueous medium. The formation of cloudy solution with micro droplets of extracting solvent dispersed in an aqueous sample allows the partitioning of target analytes from the aqueous medium to extracting solvent, which is further collected in a microsyringe for analysis. Application of ultrasound energy to DLLME (UA-DLLME) results in an increased rate of mass transfer of analytes from the aqueous phase to the extraction solvent [201]. Shi *et al.* [202] developed the UA-DLLME procedure for simultaneous analysis of multiclass herbicides, and recoveries of the analytes depending upon the type and volume of dispersion solvent, sample pH, salt concentration, amount of the sample, the ionic strength. The performance of the developed UA-DLLME was compared with that of SPE and UA-DLLME exhibited a higher enrichment factor and greater sensitivity than SPE, with limits of detection and limits of quantification of 0.004-0.024 and 0.013-0.079 µg L^{-1}, respectively, for seawater samples.

Table 13. Quantification of herbicides using different hybrid techniques.

Analyte Metabolites	Matrix	Specific Conditions	LOQ (μg/mL)	Recovery (%)	References
UAMSPD					
Ametryn, atraton, atrazine, prometon, prometryn, propazine, simetryn, terbuthylazine and terbutryne	Fruits	Ethyl acetate[a]/C8[b]/3 min[c]/40°C[d]	11.1-25.8	83-118	[203]
VAMSPD					
Atrazine and clomazone	Fish liver	ACN/C18/15 min/30°C	0.05-0.625	65-84	[204]
Atrazine, clomazone, cyhalofop and fenoxaprop	Water	Ethyl acetate/alumina and florisil/20 min/30°C	5-500	70-120	[205]
Atrazine, clomazone and fenoxaprop---ethyl	Mussel tissue	Ethyl acetate/C18/30 min/30°C	0.053-0.106	> 80	[206]
SAMSPD					
Methoxychlor and endosulfan	Tobacco	Acetone/florisil/6 hrs/40°C	1×10^{-5}-2×10^{-5}	52-77	[207]
UA-DLLME					
Chloropyrifos	Fresh vegetable	Acetone[a]/30 min[c]/30°C[d]/-undecanol[e]/1% NaCl[f]	6-24	68-88	[208]
Multiclass herbicides	Environmental waters	Chloroform/20 min/30°C/Ethanol/NaCl	0.013-0.05	> 80	[209]

[a]Extracting solvent [b]sorbent [c]Time [d]Temperature [e]dispersive solvent/salt.

DETECTION METHODS

After the extraction of herbicides from various matrices, the final step in the analytical process involves the identification and quantification of residues of herbicides quantified using high sensitivity and selectivity towards the target compound. Residues of herbicides over the years have been determined by using spectrophotometric, chromatographic, electrochemical, electrophoretic methods and biosensors.

Spectrophotometric Techniques

Spectrophotometric methods were conventionally employed for the herbicide residue estimation in various commodities. However, in recent years their use has declined to an insignificant level. This method is based on the measurement of absorbance of herbicide and often has low sensitivity and selectivity. Thus, to enhance the analyte stability and detection ability of such techniques, chemically modifying agents such as fluorophores and chromophores are added to the herbicide molecule. Some of the derivatizing agents used for herbicide determination using spectrophotometric methods are given in Table **14**.

Table 14. Derivatization reagent used for quantification of herbicide residues.

Herbicide	Matrix	Derivatizing Agent	LOD (µg/mL)	References
UV				
GLY and AMPA	Water	9-Fluorenylmethyl chloroformate (FMOC-Cl)	0.32	[210]
Fenoxaprop-p-ethyl	Wheat and barley grains	hydroxylamine hydrochloride solution	0.29	[211]
Metribuzin	Potato	p-dimethylamino-benzaldehyde	0.05	[212]
Azoxystrobin	Drug samples	hydroxylamine hydrochloride	3.8	[213]
Tebuconazole	Drug samples	ferric chloride and sodium hydroxide	0.05	[214]
Aniline	Water	triclosan	-	[215]
Oxyfluorfen	Tomato and onion	1,2-naphthoquinone-4-sulfonate	0.12	[216]
Fluorescence				
GLY and AMPA	Maize	FMOC-Cl	0.0001-0.0007	[217]
GLY, glufosinate and AMPA	Soil	FMOC- Cl	0.015-0.103	[218]
Metolachlor and buprofezin	Natural water	orthophthalaldehyde	0.026-0.0095	[219]
Chlorsulfuron, metsulfuron methyl, Rimsulfuron and sulfometuron methyl	Water	sodium dodecyl sulfate and cetyltrimethylammonium chloride	0.001-0.01	[220]
Trichloroacetic acid	Water	diphenylamine	20	[221]
Atrazine and Terbutryn	Soil and wheat grain	ammonical 2-cyanoacetamide	0.07-0.027	[222]

However, spectronic instruments have some practical limitations. These methods are relatively slow, require extensive sample preparation and cannot be used for real-time estimation. It is often difficult to extract quantitative information from signals and pure or single component samples are required. Additionally, spectrophotometer methods do not allow simultaneous analysis of multiherbicides and their TPs.

Chromatographic Techniques

TLC and HPTLC

Traditionally, TLC has been widely applied for the qualitative and quantitative analysis of herbicide residues and their transformation products from different matrices. Commercially available pre-coated silica gel plates with a florescent indicators to facilitate the detection of the compound that absorbs UV light have been widely used for the determination of herbicide residues by TLC. Alternatively, post chromatographic derivatisation with chromogenic and fluorogenic reagent, radio labelling of the analyte has been used for detection. Other sorbents reported for analysis include cellulose, alumina, kieselguhr, C_{18} RP plates impregnated with ion pairing agents such as cetyl ammonium bromide or tributyl ammonium bromide. Separation of analyte takes place through adsorption, partition, ion exchange or molecular exclusion depending on the type of relationship between the stationary phase, mobile phase and herbicide structure. Identification of herbicide is based on a comparison of the retention factor of the sample and the standard.

Though it was cost-effective method having low solvent consumption, it could not detect residues at trace levels. HPTLC is an advanced form of TLC with the advantage of automation. Due to automation, it helps to attain precision in size of sample and position at which sample is applied on TLC plate and helps to achieve gradient separation of herbicide compounds with improved sensitivity, reproducibility and accurate quantification [223]. HPTLC has been used with different detectors like conductometer and multienzyme assay. Some of the herbicides determined by TLC and HPTLC are tabulated in Table **15**.

Liquid Chromatography

Due to its high sensitivity and broad linear range, the LC method has been the widely adopted technique for the identification and quantification of polar or non-volatile and thermally labile herbicides in different matrices. Separation in LC is achieved through interaction between sample and stationary phase. The solvent and reagent used in sample preparation procedures and in HPLC analysis should not cause degradation and the unintentional reaction of analytes. They should not cause damage to the column and detector. A mixture of methanol, ACN and water in various ratios are used as eluents and separation efficiency is further increased by the addition of ammonium salt (ammonium formate/ammonium acetate), and organic acid (acetic/formic), or a combination of organic acid and ammonium salt to the mobile phase. During chromatographic analysis, the composition of the mobile phase may be kept constant ("isocratic elution mode") or varied ("gradient elution mode") during the chromatographic analysis. In gradient elution, the composition of the mobile phase is varied with a linear increase of the percentage of organic solvent. Eluting strength of the mobile phase is reflected in analyte, and faster elution is achieved when the mobile phase has high eluting strength. The common stationary phases used in herbicide residue analysis are non-polar C_8 and C_{18} [235, 236] and sorbent particle diameter generally varies from 3 to 5 μm. Short columns having particle diameter < 2 μm are gaining popularity for herbicide residue analysis. Reduced particle size necessitates the use of high pressure (UPHPLC) to achieve a sufficient eluent flow rate in the column. The use of UHPLC provides much greater efficiency as compared to conventional HPLC systems.

Detectors used for the quantification of residues fall into two main categories: universal or selective. Universal detectors measures a bulk property (*e.g.*, refractive index) by measuring the difference of physical property between the mobile phase and mobile phase with solute, while selective detectors measures a solute property by simply responding to the physical or chemical property of the solute. UV-Visible detector, diode array detector (DAD), fluorescence (FL) and mass detectors have been extensively used for the detection of herbicide residues from variable matrices. Chemical derivatization *via* the addition of suitable chromophore or fluorophore has also been widely adopted to enhance the sensitivity and selectivity of detection of herbicides lacking chromophores. Table **16** presents the various derivatization reagents used for chromatographic analysis.

Table 15. Quantification of herbicides using TLC and HPTLC.

Herbicide and Metabolites	Sample	Specific Conditions	Recovery (%)	References
TLC				
Metribuzin (P) DA, DADK and DK metribuzin (TP)	Soil	2 mm soil-water or soil[a]/biochar (1%)-water mixture[b]/Methanol/ Water[c]	70.7-105.6	[224]
Nicotine	Water	Silica based plate modified with carbon dots/Ethanol/Nicotine	94.4-105.8	[225]

(Table 15) cont.....

Herbicide and Metabolites	Sample	Specific Conditions	Recovery (%)	References
Atrazine (P) DEA and DIA	Soil	Silica/Water/Deionized water	> 85	[226]
2,4-D, 2,4,5-T, MCPA, 2,4-DP, dicamba, MCPP and MCPB (P)	Water	silica gel 60 F_{254} aluminium plate/ACN/20% sulphuric acid	86.1-92.6	[227]
Atrazine, simazine and propazine (P)	Soil	silica gel 60/methanol/toluene-acetone mixture	> 90	[228]
HPTLC				
Atraton, terbumeton, simazine, atrazine and terbuthylazine (P)	Water	Silica gel Licrospere plate with fluorescent dye/Methyl t-buthyl ether cyclohexane/Chlorine	> 80	[229]
Profenofos	Visceral	silica gel 60 F_{254}/hexane:acetone (7:3,v/v)/cupric acetate reagent	> 80	[230]
2,4-D, 2,4-DB and 2,4-DP	Water	silica gel 60 F_{254}/ Methanol/Nitrogen	> 85	[231]
Atrazine (P)	Soil	silica gel 60 F_{254}/Methanol/Nitrogen	98.7-103.5	[232]
2,4-D, 4-(2,4)-DB and 2,4,5-T	Water	Silica gel with UV-254-indicator/ACN/Toluene	> 80	[233]
Bensulfuron methyl	Soil	silica gel 60 F_{254}/ dichloromethane-aceton--methanol-aqueous ammonia (45:15:10:1)	>80	[234]

aCoating of TLC plate bDeveloping solvent cStaining solution.

Table 16. Derivatization reagent used for chromatographic analysis of residues of the herbicide.

Detector	Herbicide	Matrix	Derivatizing Agent	LOD (µg/mL)	References
HPLC					
UV	Amitrole GLY	Water	4-chloro-3,5-dinitrobenzenotrifluoride 9- fluorenylmethylchoroformate o-nitrobenzenenesulfonyl chloride, p-toluenesulphonyl chloride	0.16, 0.00005, -0.005, 6.25, 0.02	[237 - 242]
PDA	Chlorophenoxy acid	Millet and pepper	2-(11H-benzo[a]carbazol-11-yl)-ethyl-4-methylbenzenesulfonate	0.00046-0.00079	[243]
DAD	Amitrole	Apple	4-Chloro-3,5-dinitrobenzotrifluoride	0.10	[237]
GC					
NPD	Glyphosate (GLY)	Soil	Trifluricacetic anhydride and 4,4,4-trifluoro-1-butanol	0.02	[244]

(Table 16) cont.....

Detector	Herbicide	Matrix	Derivatizing Agent	LOD (µg/mL)	References
NPD	GLY	Water	Proppionoic anhydruide	5×10^{-7}	[245]
FID	GLY	Water	N-methyl-N-tert-butyldimethylsiliconcontrifluoroacetamide	-	[246]
FID	GLY	Water	Trifluoroacetic acid- trifluoroacetic anhydride and trimethyl orthoformate	-	[247]
ECD	GLY	Water	Trifluoroacetic anhydride and 4,4,4-trifluoro-1-butanol	1.0	[248]

Mass spectrometry is a key analytical technique that enables the quantification and conformation of herbicide molecules. It measures the mass-to-charge ratio of the charged molecule. An ion source, analyser and detector are three basic components of the mass spectrometer. Atmospheric pressure (AP), electrospray ionization (ESI) and chemical ionization (APCI) are frequently used ionization methods for the detection and quantification of herbicides of various chemical classes from complex matrices. Analyser separates the ions on the basis of mass to charge ratio (m/z) and a direct beam of focussed ions to the detector. Different kinds of analysers vary in terms of characteristics and operation, and quadrupole (Q), time of flight (TOF) and ion trap (IT) mass analysers are frequently used analysers for herbicide residue analysis. The separation efficiency of the quadrupole analyser is a function of ion mass and increases slightly with an increase in mass. TOF separates ions accelerated by an electric field according to their velocity, which depends on their m/z values. It measures the time ions have a specific m/z ratio to reach the detector, and the smaller is m/z ratio faster the ions reach the detector. TOF analyzer is characterized by a broad measuring range, high scanning speed and high sensitivity. In herbicide residue analysis, TOF has been used to confirm the presence of herbicides in various matrices such as fruit, vegetables and honey. IT analyzer traps ion with a specific m/z ratio by an electric field and send them to the detector in order to increase m/z values. Table **17** summarises various LC methods for the extraction of herbicides and their transformation products from various matrices.

Gas Chromatography (GC)

GC is used for separating and analyzing compounds that can be vaporized without decomposition. Separation of the analyte occurs because of its partition of analyte between stationary and mobile phase. In GC, the mobile phase is a carrier gas, usually an inert gas such as helium or an unreactive gas such as nitrogen. Helium is the commonly used carrier gas in about 90% of instruments, although hydrogen is preferred for improved separations. The stationary phase is a microscopic layer of liquid or polymer on an inert solid support, inside a piece of glass or metal tubing called a column. The separation column can be divided into two distinct groups, *i.e.*, packed and capillary columns of various dimensions. As better separation capacity and resolution of peaks are obtained with the capillary column, these have been more often used for quantification of herbicide residues. The sample component with large distribution coefficients is more strongly adsorbed/retained in the stationary phase and takes a longer time to pass through the column.

The mobile phase used in GC is an inert gas such as nitrogen, helium or hydrogen, referred to as carrier gas. The nature of carrier gas may influence the separation characteristics and sensitivity of detection. Operation of GC is performed using set temperature or by systematically increasing temperature during the analysis. To be amenable to GC analysis, compounds must have a relatively high vapour pressure. Many herbicides have been analysed in their native form, while many other have been derivatised to make them sufficiently volatile, less polar and/or more thermally stable. Phenoxy acid herbicides have strong polarity and high boiling point, are not volatile and are derivatized with methanol, pentafluorobenzyl bromide (PFBBr) and diazomethane before GC analysis (Table 16).

Table 17. Quantification of herbicides using different LC techniques.

Analyte	Column	Mobile Phase	LOD (µg/mL)	References
HPLC-UV				
2,4-D, atrazine and alachlor	H5-ODS C18[a], 15 cm × 4.6 mm[b], 5 µm[c]	water/ACN (30:70 v/v)	5×10^{-5}-0.0001	[249]
Simazine, atrazine, prometon, ametryn, propazine and terbutryn	C_{18}, 3.9 × 300 mm, 4 µm	25 mM monopotassium phosphate:ACN (60:40,*v/v*)	0.19-0.68	[250]
Imazaquin	C18, 4.6 ×250.0 mm, 2 µm	Water, ACN, with phosphoric acid	5	[251]
HPLC-DAD				
Simazine, atrazine, prometon, ametryn and prometryn	C18, 4.6 × 250 mm, 5.0 µm	methanol-water (75:25, v/v)	5.3×10^{-5}-0.0002	[252]
Ametryn, atrazine, cyanazine, prometryn, propazine, simazine, simetryn, terbuthylazine and terbutryn	C18, 150mm × 4.6mm, 5 µm	ACN:H_2O gradient elution	0.007-0.235	[253]
Atrazine	C18, 145mm × 4.7mm, 5 µm	0.05% trifluoroacetic acid (TFA) in water,water and 0.05% TFA in ACN, in gradient elution	0.0003-0.001	[254]

(Table 17) cont.....

Analyte	Column	Mobile Phase	LOD (µg/mL)	References
metsulfuron-methyl, chlorsulfuron, chlorotoluron, and bensulfuron-methyl	C18, 4.6 ×250 mm, 5 µm	methanol-0.05% acetic acid solution (50:50, v/v)	0.02-0.07	[255]
HPLC-MS				
Azimsulfuron, bensulfuron methyl and mesotrione	C_{18}, 100 × 4.6 mm, 2.6 µm, Quattro Micro QQQ, ESI-	0.1% formic acid in MeCN:0.1% formic acid in water	0.0003-0.015	[256]
Rimsulfuron, mesotrione, fluroxypyr-meptyl and fluroxypyr	Zorbax SB-C_{18} column (50×2.1 mm id, 1.8 µm), Quattro Micro QQQ, ESI+	0.1% formic acid and ACN	-	[257]

[a]Type of column, [b]dimensions of column, [c]particle size, [d]analyser, [e]ion source.

GC equipped with detectors like FID, NPD, ECD, and mass spectrometer has been used for sensitive and selective detection and quantification of herbicides (Table **18**). Mass spectrophotometer detectors have replaced traditional detectors like FID, and ECD in newly developed methods, due to high sensitivity, selectivity and low limit of detection. In GC/MS system, the mass spectrometer scans the mass continuously throughout the separation. The sample is ionised and fragmented by an electron impact ion source. During this process, a sample is bombarded by energetic electrons, which ionises the molecule, causing it to lose electrons by electrostatic repulsion. Further bombardment causes the ions to be fragmented. The ions are then passed into a mass analyser where they are sorted according to their mass to charge value.

Several ionisation techniques have been used in GC-MS over the years. Amongst these, electron ionisation (EI) is most widely used as it is a robust and universal ionisation source that generates highly reproducible spectra. However, in some cases, due to highly extensive fragmentation, sensitivity is very poor. In these cases, chemical ionisation has been used to obtain a predominant molecular ion peak. A promising source is APCI. The most common type of mass analyser is ion trap mass, quadrupole ion trap and TOF analyser. TOF has characteristics of a broad range of measurement, high sensitivity and high scanning speed. While the separation efficiency and m/z range of the ion trap are similar to that of the quadrupole ion trap.

Electrochemical Methods

Electrochemical techniques constitute an alternative to chromatographic and spectroscopic methods and have received much attention due to their high selectivity and sensitivity. Electrochemical techniques have advantages over other techniques related to their simplicity, portability, low sample volumes required (typically in the order of μL), low cost of instruments and short analysis time.

Table 18. Quantification of herbicides residues using GC from various matrices.

Analyte	Column	LOD (μg/mL)	References
GC-ECD			
Acetochlor, alachlor, metolachlor, butachlor and pretilachlor	KB-5 fused silica capillary,30 m×0.25mm× 0.25 mm, 0.5 μm	0.000006-0.0001	[258]
Butanone	Fused silica, 30 × 0.25 mm, 0.25 μm	-	[259]
Chlorpyrifos-ethyl	silica column, 25 ×0.530 mm, 0.25 μm	0.001	[260]
trifluralin, atrazine, acetochlor, and alachlor	capillary column, 20 × 0.15 mm, 0.25 μm	0.004-0.0011	[261]
GC-NPD			
Metobromuron, monolinuron, linuron	BP10, 30m×0.25mm×0.25m, 0.25 μm	0.03-0.32	[262]
Diazinon, atrazine, simazine, dimethoate, and prometryn	Rtx-1701 column, 20 × 0.25 mm, 0.25 μm	0.01	[263]
Terbuthylazine, simazine, and atrazine	HP-5 capillary column, 40 × 0.35 mm, 0.25 μm	0.001-0.007	[264]
GCMS			
MCPB, MCPP and triclopyr	Fused silica capillary column, 30 m × 0.25 mm, 0.25 mm, quadrupole mass selective[d], ESI+[e]	0.0018-0.003	[265]
Simazine, atrazine, propazine, secbumeton, sebuthylazine, desmetryn, simetryn and prometryn	DB-5 column, 30 m, 0.32-mm, 0.25, quadrupole mass selective detector, ESI	0.007-0.063	[266]
Ametryn, atraton, atrazine, prometon, prometryn	BPX5 column, 30m×0.25mm, 0.25, quadrupole mass selective, ESI+	0.003-0.008	[267]
Pendimethalin, Diazinon, malathion, prometryn	capillary columnVF-5, 60 m × 0.25 mm × 0.25, 0.25 μm, quadrupole mass selective, ESI+	0.0005-0.002	[268]
Malathion, diazinon and 1,4-dichlorobenzene	capillary column, 30 m ·0.25 mm, 0.25 μm, quadrupole mass selective, ESI+	0.0005-0.003	[269]

[a]Type of column, [b]dimensions of column, [c]particle size, [d]analyser, [e]ion source.

Electrochemical methods are the analytical techniques that use a measurement of potential, charge or current to determine an analyte's concentration or to characterise an analyte's chemical reactivity. Commonly used electrochemical techniques for analysis of herbicide residues in water, soil and food include voltammetric, potentiometric, amperometric and conductometry.

Electrochemical techniques have mostly been used for nitro or other electrochemically active substituent herbicides. However, the selection of electrochemical method and LOD depends upon the nature of the sample and analyte (Table **19**).

Table 19. Electrochemical techniques for quantification of herbicides.

Herbicide	Electrode	Technique	Matrix	Electrolyte	References
Alachlor	SMDE[a]	Dpv[b]	Soil and water	Phosphate buffer	[270]
Atrazine	DME[c]	DPV	Water	KCl/HCl buffers	[271]
Diquat	SMDE	SMV[d]	Potatoes	NaOH	[272]
Paraquat	SMDE	SMDE	Potatoes	NaOH	[273]
Trifluralin, Benfluralin, Pendimethalin	Dme, SMDE	AdSV	Soil	Britton-robinson buffer	[274]
Propanil	Glassy carbon	CV[f], DPV, SWV	-	Britton-robinson buffer	[275]
s-triazine Sulphonylureas	HMDE Calomel	DPP DPV	Wastewater -	0.1M H_3PO_4 Mercury (II) solutions in 0.1 M KNO_3	[276, 277]
Paraquat	Chitin modified	SWASV[g]	Olive	Na_2SO_4	[278]
Metribuzin	Glassy carbon	CV, DPV SWV	-	20% v/v ethanol/ phosphate buffer	[279]
Chlorsulfuron	Screen printed	Amperometry	Water	Ag/AgCl	[280]
Diuron and fenuron	Glassy carbon	Amperometry	Water	0.1 M phosphate buffer	[281]

[a]SMDE: static mercury drop electrode; [b]DPV: Differential pulse voltametry; [c]DME: dropping mercury electrode; [d]SWV: Square-wave voltametry; [e]AdSV: Adsorptive square-wave voltametry; [f]CV: Cyclic voltametry; [g]SWASV: Square wave anodic stripping voltametry

The performance of electrochemical method is dependent on material of working electrode and its selection is based on the redox behaviour of target analyte and the background current over the applied potential range. Traditionally dropping mercury electrode (DME), static mercury drop electrode (SMDE) and hanging mercury electrode (HMDE) has been used for detection of herbicide residues in various matrices.

In order to remove the drawbacks of the classical bare electrode (mercury electrode), such as adsorption of molecules or ions, unpredictable surface reactivity and sluggish kinetics, efforts have been made to alter the electrodes configuration by the chemical modification, *viz.* carbon nanotube modified electrode, carbon paste electrode, clay-film modified electrode, immobilized enzyme electrode and miniaturization.

Electrophoretic Methods

Capillary electrophoresis (CE) is an electrophoretic technique that separates ions based on their electrophoretic mobility with the use of an applied voltage. The electrophoretic mobility is dependent upon the charge of the molecule, the viscosity, and the atom's radius. The rate at which the particle moves is directly proportional to the applied electric field; the greater the field strength, the faster the mobility. Because of its high separation efficiency, versatility, and rapid qualitative and quantitative analysis, CE is a good alternative to widely used chromatographic techniques like HPLC and GC.

Various types of CE techniques *viz.* capillary zone electrophoresis (CZE) and Micellar electrokinetic chromatography (MEKC), have been used. CZE is based on the differences in m/z ratios of the analytes during electrophoresis. CZE has been used with various detectors *viz.* UV, capacitively coupled contactless conductivity detector (C^4D), diode array detector (DAD), laser-induced fluorescence (LIF), enhanced chemiluminescence (ECL), and MS. MEKC has been successfully used for neutral and ionic herbicides (Tables **20**) and involves differential partitioning between micellar and aqueous phases.

Table 20. CE techniques for quantification of herbicide residues [282 - 285].

Analytes	CE Mode	Matrix	Detection	Separation Buffer	LOD
GLY (P) AMPA (TP)	CZE	water	DAD	25 mM tetraborate	0.4 ng/mL
GLY (P), glufosinate, bialaphos, AMPA, 3-methylphosphinicopropionic acid (TP)	CZE	Soil, tea	MS	100 mM formic acid adjusted with 100 mM ammonia	0.5-10 µg/mL
GLY (P) AMPA, glyoxylate (TP)	CZE	Lolium spp.	DAD	0.5 mM CTAB, 10mM potassium phthalate, 10% ACN	0.1-0.2 µg/mL
GLY (P) glufosinate,AMPA (TP)	CZE	Water and soil	LIF	15 mM Brij-35,30 mM boric acid	1.99-6.14 ng/kg
GLY (P), AMPA (TP)	MEKC	Tap and river water	C^4D	75 µM CTAB, 12mM histidine, 8 mM MES, 3% methanol	0.06-0.005 µg/L
GLY (P), glufosinate-ammonium,AMPA (TP)	MEKC	Apple surface	LIF	10 mM SDS, 10 mM sodium tetraborate, 10% (v/v) ACN	1-10 ppb

(Table 20) cont.....

Analytes	CE Mode	Matrix	Detection	Separation Buffer	LOD
Diquat, Paraquat, difenzoquat (P)	CZE	Tap and river water	UV	150 mM phosphate (pH 2.4)	0.5 ng/mL
Diquat Paraquat, (P)	CZE	Tap water	DAD	10% ethanol, 50 mM 1-butyl-3-methylimidazolium hexafluorophophate	1.95-2.59 µg/L
Linuron, Isoproturon, diuron (P)	CE	Vegetables, rice	ECL	12 mg/mL poly-β-CD, 20 mM phosphate	0.1-0.2 µg/L
Monolinuron Monuron,, diuron (P)	CZE	yam	ECL	25 mM phosphate	0.01-0.05 µg/L
Halosulfuron-methyl (P)	CZE	Sugarcane juice, tomato	MS	20 mM NH_4HCO_3	2 ppb
Chlorsulfuron, Metsulfuron methyl (P)	CZE	Lake, reservoir, underground water	DAD	50 mM tetraborate, 3% methanol	0.36-0.40 µg/L
3,6-dichloro-2-methoxybenzoic acid, 2,4-dichlorophenoxybutyric acid, 2,4-dichlorophenoxyacetic acid (TP)	CE	River water	UV	100 mM phosphate, 1 mM α-CD	10-15 ng/mL
Simazine Atrazine, ametryn prometryn, terbutryn (P)	MEKC	Well, river, reservoir water	UV	10 mM borate, 2.5% SDS, 0.8% ethyl acetate, 6.0% 1-butanol	0.41-0.62 ng/mL

CE is advantageous in terms of short analysis time, low reagent consumption, high separation efficiency and low operational cost. However, poor sensitivity and low sample injection volumes significantly limit CE application to analyses of herbicides. However, the on-line focus has been explored for the determination of herbicide from various metrics (Table **21**). Various on-line focusing strategies are field-amplified sample injection (FASI), large-volume sample stacking (LVSS), analyte focusing by micelle collapse (AFMC), dynamic pH junction, transient isotachophoresis and sweeping.

Table 21. Online focusing strategies for the determination of herbicides by CE [282].

Herbicide	Matrix	CE method	Detection	LOD
Atrazine, ametryn, atraton (P) 2-hydroxyatrazine (TP)	Water	pH-mediated-CZE	UV	0.054-0.31 µM
GLY (P), glufosinate, AMPA (TP)	Tap water	LVSS-CE FESI-CE	C^4D	0.1-2.2 µg/L

(Table 21) cont.....

Herbicide	Matrix	CE method	Detection	LOD
Triasulfuron, rimsulfuron, flazasulfuron, metsulfuron-methyl, chlorsulfuron (P)	Ground water	LVSS-CZE	DAD	0.04-0.12 µg/L
Nicosulfuron, tribenuron methyl, thifensulfuon methyl, sulfometuron methyl, chlorimuron ethyl, pyrazosulfuron ethyl (P)	Rice, wholemeal, flour oatmeal	LVSS-MEKC	UV	0.22-0.89 ng/g
Simazine, prometryn, ametryn, atrazine, propazine, simetryn, terbuthylazine, terbutryn (P)	Cereal, carrots, chives	Sweeping-MEKC	DAD	0.02-0.04 ng/g
bensulfuron methyl, chlorsulfuron, chlorimuron ethyl, tribenuron methyl, metsulfuron methy (P)	Soil	Sweeping-MEKC	DAD	0.5-1. 0 ng/g
Diquat, parathion, paraquat, fenitrothion, difenzoquat, azinphos-methyl (P)	Water	AFMC (Sweeping)	UV	0.004-0.02 µg/mL

C⁴D: capacitively coupled contactless conductivity detection; MEKC: micellar electrokinetic chromatography AFMC: analyte focusing by micelle collapse; CZE: capillary zone electrophoresis; CE: capillary electrophoresis; LVSS: large volume sample stacking; DAD: diode array detector;

Hyphenated Techniques

During the last decade, in order to enhance the sensitivity, selectivity and separation efficiency of the measurement and to lower the detection limit, hyphenation, *i.e.*, a combination of two or more techniques, has been successfully exploited for the identification and quantification of complex compounds in short duration. LC-MS/MS and GC-MS/MS are the most popular hyphenated techniques used for herbicide residue analysis. Tandem mass spectrometry (MS/MS) is a system of two combined analyzers of the same type or different types, characterized by high separation efficiency, sensitivity and selectivity as compared to the single analyser. The main possibilities for connecting different types of analysers include triple quadrupole systems (QQQ), quadrupole time-o--flight systems (Q-TOF), and quadrupole-linear ion trap systems (Q-Trap). In comparison with analysis using a single analyzer, tandem analysis involves improved selectivity and considerably increased sensitivity. The hyphenation also involves more than one separation or detection technique, *e.g.*, LC-PDA-MS, LC-MS-MS, LC-NMR-MS, LCPDA-NMR-MS, *etc.* The main aim of this coupling is to obtain an information-rich detection for both identification and quantification as compared to that of a simple detector for any system. Table **22** summarises various hyphenated techniques used for the analysis of herbicide residues.

Table 22. Quantification of herbicides using hyphenated techniques.

Analyte	Analyser/Ion Source	Column	Mobile Phase	LOD (µg/mL)	References
HPLC-MS/MS					
Ametryn, atrazine, desmetryn, prometryn, propazine and simazine	Q-Trap MS/ API	C_{18}, 250 × 4.6 mm, 5 um	ACN/water (68:32)	0.003-0.0065	[286]
Ethametsulfuron-methyl, pyrazosulfuron-ethyl, chlorimuron-ethyl, cyclosulfamuron	Q-Trap/API	Zorbax SB-C18, 225 × 3.4 mm, 5 um	0.1% formic acid (FA) and MeCN	0.0004- 0.002	[287]
GLY and AMPA	Q-Trap/API	C18, 235 × 4.6 mm, 2 um	FA in MeCN, and FA in water	0.05	[288]
UPLC-MS/MS					
Atrazine, phenmedipham, mefenacet, metholachlor and hexazinone	QQQ mass/APCI	BEH C18, 50 × 2.1 mm, 1.7 µm	Methanol and water	0.00005-0.0002	[289]
Phenylureas and triazines	triple quadrupole mass spectrometer/API	BEH C18, 2.1 × 50 mm, 1.7 µm	H_2O with 0.05% FA, MeOH with 0.05% FA	0.0005-0.002	[290]
Multi herbicides	Q- Orbitrap-MS/ESI	Q-MS column, 100 mm × 2.1 mm, 2.6 µm	ACN and water acidified with acetic acid	0.001	[291]
Pyroxsulam, flumetsulam, metosulam, and diclosulam	QQQ mass spectrometer/API	UPLC BEH C18, 100 mm × 2.1 mm, 3.6 µm	ACN, and 0.1% FA in water	0.001-0.006	[292]
LC-MS/MS					
Four phenylureas, five SUs	TOF-MS/APCI	XDB-C18, 100 mm × 2.1 mm, 4.6 µm	ACN containing 0.1% FA	0.3-9.7	[293]
Atraton, Simetryn, Prometon, Ametryn, Propazin and Prometryn	QQQ mass spectrometer/APCI	C18 column, 100 mm × 2.1 mm, 3.9 µm	methanol-water (85:15)	0.000001-0.000004	[294]
Ametryn, Prometryn, Simazine and Simetryn	ion trap system/ESI	Waters Xterra MS C_{18}, 150 mm 4.6 mm), 2.5 µm	ACN containing 0.1% FA: water containing 0.1% FA	0.01	[295]

(Table 22) cont.....

Analyte	Analyser/Ion Source	Column	Mobile Phase	LOD (µg/mL)	References
3-aminopropyltriethoxysilane (APTES), chlorimuron-ethyl, nicosulfuron, thifensulfuron-methyl and chlorsulfuron	triple quadrupole mass spectrometer/ESI	C_{18} BEH, 50× 2.1 mm, 1.7 µm)	mixture of 0.1% methane acid and ACN	0.000006-0.001	[296]
Atrazine, chlorimuron ethyl, metolachlor, metsulfuron methyl and nicosulfuron	QTOF-MS/APCI	BEH C18 (100 mm×, 2.1mm, 1.7µm	ACN containing 0.1% FA	0.00006	[297]
Diuron, monuron, and monolinuron	QIT-MS	Luna C18, 200×2.1mm, 2.4 µm	ACN containing 0.1% FA	0.005-0.0010	[298]
GC-MS/MS					
Atrazine, propazine, simazine, diuron, bromophos terbuthylazine, oxyfluorfen, and norflurazon	Ion-Trap MS/MS detector/Electron impact ionisation	Varian CPSil8CB	aqueous 0.1% FA and Methanol	0.0005-0.003	[299]
Chlorinated herbicides	QTOF-MS	capillary column, 30 × 0.25 mm × 0.25 µm	ACN:water (8:2)	0.03-7.1	[300]

Biosensors

Considering the growing concern regarding the monitoring of herbicide residues, the development of miniaturized and disposable tools called biosensors having high stability, selectivity accuracy, and low detection limit is a growing concern. Biosensors are defined as analytical devices which convert a biological response to a quantifiable and processable signal. Two main groups of biosensors can be identified: (1) those dealing with direct measurement of the analyte or a product of an enzymatic reaction and (2) those dealing with indirect quantification of the analyte evaluating the decrease in the electrochemical signal of the biosensor caused by the poisoning the sensor element when the analyte is present.

Depending upon biorecognition elements, different biosensors include cell-based, enzyme-based, immunosensors, and nucleic acid-based biosensors having voltammetric, amperometric or potentiometric as transducers. Due to extremely high sensitivity and selectivity, electrochemical sensors have attracted much attention over the last two decades for the determination of herbicide residues. Biosensors have been developed for the detection of various herbicides, such as atrazine, diuron, 2,4-D, paraquat, isoproturon, propanil, *etc.* (Table **23**).

Biosensors help to perform on-site analysis and predict the extent of pollution almost immediately. The use of biosensors is advantageous as these are reliable, disposable, selective and economical and can be miniaturized for efficient use. These require less sample size and are easy to operate. However, biosensors have low response stability, low mechanical stability, high diffusion resistance of substrate/bio component assembly and interfering signals from other compounds in real samples affect the prediction of herbicide residues in real samples. These drawbacks can be overcome or minimized by proper designing of the biosensor.

Nanomaterials have enabled advances in sensor design such as miniaturization, portability, and rapid signal response times. High surface area to volume ratios and facile surface functionalization make nanomaterials highly sensitive to changes in surface chemistry, thus enabling nanosensors to achieve extremely low detection limits. In some cases, the enhanced sensitivity of nanoenabled sensors is due to the fact that nanomaterials are of a similar size as the analyte of interest (*e.g.*, metal ions, pathogens, biomolecules, antibodies, DNA) and are thus capable of interrogating previously unreachable matrices.

Table 23. Biosensor detection of residues of herbicides.

Herbicide	Biosensor	Electrode	Transducer	References
Atrazine, simazine and diuron	*Synechococcus elongatus*	Ag/Pd	Clark electrode	[301]
Atrazine	Tyrosinase	Gold	Amperometric	[302]
Atrazine	Tyrosinase	Poly 1-tyrosinase	Amperometric	[303]
Atrazine	Thylakoids	Graphite	Amperometric	[304]
Bispyribac sodium	Polyvinyl chloride membrane based biosensor	Screen printed carbon electrode	Potentiometric	[305]
Diuron	Thylakoid membrane	Graphite	Amperometric	[306]
Diuron	Fabricated by simple electrochemical co-reduction of graphene oxide and chloroauric acid	Graphene oxide-gold nanoparticle	Cyclic voltammetry	[307]
3-(3,4-dichloropehnyl)1,1-dimethylurea	Cynobacterium	Gold disc	Voltametric	[308]
Chlortoluron	Tyrosinase	Modified screen printed carbon electrode	Voltamperometric	[309]
Diuron and fenuron	Graphene oxide-MWCNT (GO-MWCNT) film	Modified glassy carbon electrode	Amperometric	[310]
3-(3,4-dichloropehnyl)1,1-dimethylurea	*Nostoc muscorum*	Gold disc	Differential pulse voltammetry	[311]

Herbicide	Biosensor	Electrode	Transducer	References
GLY	Horseradish peroxidase (HRP)/ g poly(2,5-dimethoxyaniline) (PDMA)/poly(4-styrenesulfonic acid) (PSS) based biosensor	Gold	Voltammetric and amperometric	[312]
Atrazine, isoproturon and propanil	*Chlorellauulgaris*	Carbon cathode oxygen	Amperometric	[313]

REFERENCES

[1]   Food and Agriculture Organization of the United Nations **2018**. Available at: http://www.fao.org/faostat/en/

[2]   Jabusch, T.W.; Tjeerdema, R.S. Microbial degradation of penoxsulam in flooded rice field soils. *J. Agric. Food Chem.,* **2006**, *54*(16), 5962-5967.
[http://dx.doi.org/10.1021/jf0606454] [PMID: 16881702]

[3]   Kalsi, N.K.; Kaur, P. Dissipation of bispyribac sodium in aridisols: Impact of soil type, moisture and temperature. *Ecotoxicol. Environ. Saf.,* **2019**, *170*, 375-382.
[http://dx.doi.org/10.1016/j.ecoenv.2018.12.005] [PMID: 30550967]

[4]   Suzuki, T.; Casida, J.E. Metabolites of diuron, linuron, and methazole formed by liver microsomal enzymes and spinach plants. *J. Agric. Food Chem.,* **1981**, *29*(5), 1027-1033.
[http://dx.doi.org/10.1021/jf00107a035] [PMID: 7309986]

[5]   Chen, H.; Zhang, Z.; Feng, M.; Liu, W.; Wang, W.; Yang, Q.; Hu, Y. Degradation of 2,4-dichlorophenoxyacetic acid in water by persulfate activated with FeS (mackinawite). *Chem. Eng. J.,* **2017**, *313*, 498-507.
[http://dx.doi.org/10.1016/j.cej.2016.12.075]

[6]   Ramakrishna, M.; Venkata Mohan, S.; Shailaja, S.; Narashima, R.; Sarma, P.N. Identification of metabolites during biodegradation of pendimethalin in bioslurry reactor. *J. Hazard. Mater.,* **2008**, *151*(2-3), 658-661.
[http://dx.doi.org/10.1016/j.jhazmat.2007.06.039] [PMID: 17683860]

[7]   Amaral, B.; de Araujo, J.A.; Peralta-Zamora, P.G.; Nagata, N. Simultaneous determination of atrazine and metabolites (DIA and DEA) in natural water by multivariate electronic spectroscopy. *Microchem. J.,* **2014**, *117*, 262-267.
[http://dx.doi.org/10.1016/j.microc.2014.07.008]

[8]   Ibáñez, M.; Sancho, J.V.; Pozo, Ó.J.; Hernández, F. Use of quadrupole time-of-flight mass spectrometry in environmental analysis: elucidation of transformation products of triazine herbicides in water after UV exposure. *Anal. Chem.,* **2004**, *76*(5), 1328-1335.
[http://dx.doi.org/10.1021/ac035200i] [PMID: 14987089]

[9]   Kaur, R.; Goyal, D. Biodegradation of Butachlor by *Bacillus altitudinis* and Identification of Metabolites. *Curr. Microbiol.,* **2020**, *77*(10), 2602-2612.
[http://dx.doi.org/10.1007/s00284-020-02031-1] [PMID: 32435881]

[10]   Kaur, P.; Kaur, P. Time and temperature dependent adsorption-desorption behaviour of pretilachlor in soil. *Ecotoxicol. Environ. Saf.,* **2018**, *161*, 145-155.
[http://dx.doi.org/10.1016/j.ecoenv.2018.05.081] [PMID: 29879575]

[11]   López-Ruiz, R.; Romero-González, R.; Martínez Vidal, J.L.; Fernández-Pérez, M.; Garrido Frenich, A. Degradation studies of quizalofop-p and related compounds in soils using liquid chromatography coupled to low and high resolution mass analyzers. *Sci. Total Environ.,* **2017**, *607-608*, 204-213.
[http://dx.doi.org/10.1016/j.scitotenv.2017.06.261] [PMID: 28692891]

[12]   Wu, J.; Wang, K.; Zhang, Y.; Zhang, H. Determination and study on dissipation and residue determination of cyhalofop-butyl and its metabolite using HPLC-MS/MS in a rice ecosystem. *Environ. Monit. Assess.,* **2014**, *186*(10), 6959-6967.
[http://dx.doi.org/10.1007/s10661-014-3902-7] [PMID: 25007772]

[13]   Guan, W.; Ma, Y.; Zhang, H. Dissipation of clodinafop-propargyl and its metabolite in wheat field ecosystem. *Bull. Environ. Contam. Toxicol.,* **2013**, *90*(6), 750-755.
[http://dx.doi.org/10.1007/s00128-013-0997-4] [PMID: 23612716]

[14]   Harir, M.; Frommberger, M.; Gaspar, A.; Martens, D.; Kettrup, A.; El Azzouzi, M.; Schmitt-Kopplin, P. Characterization of imazamox degradation by-products by using liquid chromatography mass spectrometry and high-resolution Fourier transform ion cyclotron resonance mass spectrometry. *Anal. Bioanal. Chem.,* **2007**, *389*(5), 1459-1467.
[http://dx.doi.org/10.1007/s00216-007-1343-7] [PMID: 17554530]

[15]   Vega, D.; Cambon, J.P.; Bastide, J. Triflusulfuron-methyl dissipation in water and soil. *J. Agric. Food Chem.,* **2000**, *48*(8), 3733-3737.
[http://dx.doi.org/10.1021/jf9910943] [PMID: 10956179]

[16]   Hultgren, R.P.; Hudson, R.J.M.; Sims, G.K. Effects of soil pH and soil water content on prosulfuron dissipation. *J. Agric. Food Chem.,* **2002**, *50*(11), 3236-3243.
[http://dx.doi.org/10.1021/jf011477c] [PMID: 12009993]

[17]   Jackson, R.; Ghosh, D.; Paterson, G. The soil degradation of the herbicide florasulam. *Pest Manag. Sci.,* **2000**, *56*(12), 1065-1072.
[http://dx.doi.org/10.1002/1526-4998(200012)56:12<1065::AID-PS252>3.0.CO;2-Y]

[18]   Zhao, W.; Wang, C.; Xu, L.; Zhao, C.; Liang, H.; Qiu, L. Biodegradation of nicosulfuron by a novel Alcaligenes faecalis strain ZWS11. *J. Environ. Sci. (China),* **2015**, *35*, 151-162.
[http://dx.doi.org/10.1016/j.jes.2015.03.022] [PMID: 26354704]

[19]   Sweetser, P.B.; Schow, G.S.; Hutchison, J.M. Metabolism of chlorsulfuron by plants: Biological basis for selectivity of a new herbicide for cereals. *Pestic. Biochem. Physiol.,* **1982**, *17*(1), 18-23.
[http://dx.doi.org/10.1016/0048-3575(82)90121-3]

[20]   Saha, S.; Kulshrestha, G. Degradation of sulfosulfuron, a sulfonylurea herbicide, as influenced by abiotic factors. *J. Agric. Food Chem.,* **2002**, *50*(16), 4572-4575.
[http://dx.doi.org/10.1021/jf0116653] [PMID: 12137477]

[21]   Choudhury, P.P.; Dureja, P. Phototransformation of Chlorimuron-ethyl in Aqueous Solution. *J. Agric. Food Chem.,* **1996**, *44*(10), 3379-3382.
[http://dx.doi.org/10.1021/jf950556j]

[22]   Yasuor, H.; Zou, W.; Tolstikov, V.V.; Tjeerdema, R.S.; Fischer, A.J. Differential oxidative metabolism and 5-ketoclomazone accumulation are involved in Echinochloa phyllopogon resistance to clomazone. *Plant Physiol.,* **2010**, *153*(1), 319-326.
[http://dx.doi.org/10.1104/pp.110.153296] [PMID: 20207709]

[23]   Rice, P.J.; Anderson, T.A.; Coats, J.R. Degradation and persistence of metolachlor in soil: Effects of concentration, soil moisture, soil depth, and sterilization. *Environ. Toxicol. Chem.,* **2002**, *21*(12), 2640-2648.
[http://dx.doi.org/10.1002/etc.5620211216] [PMID: 12463559]

[24]   Barchanska, H.; Rusek, M.; Szatkowska, A. New procedures for simultaneous determination of mesotrione and atrazine in water and soil. Comparison of the degradation processes of mesotrione and atrazine. *Environ. Monit. Assess.,* **2012**, *184*(1), 321-334.
[http://dx.doi.org/10.1007/s10661-011-1970-5] [PMID: 21416215]

[25]   Elmarakby, S.A.; Supplee, D.; Cook, R. Degradation of [(14)C]carfentrazone-ethyl under aerobic aquatic conditions. *J. Agric. Food Chem.,* **2001**, *49*(11), 5285-5293.
[http://dx.doi.org/10.1021/jf010601p] [PMID: 11714318]

[26]   Xu, H.; Pan, W.; Song, D.; Yang, G. Development of an improved liquid phase microextraction technique and its application in the analysis of flumetsulam and its two analogous herbicides in soil. *J. Agric. Food Chem.,* **2007**, *55*(23), 9351-9356.
[http://dx.doi.org/10.1021/jf0718345] [PMID: 17953444]

[27]   Kaur, P.; Kaur, P.; Bhullar, M.S. Environmental aspects of herbicides use under intensive agriculture scenario of Punjab. In: *Shobha Sondia*; Choudhary, P.P.; Sharma, A.R., Eds.; Herbicide Residue Research in India. A Springer Publication, **2019**; pp. 101-157.
[http://dx.doi.org/10.1007/978-981-13-1038-6_3]

[28]   Anonymous The bioassay technique in the study of the herbicide effects.Herbicides, Theory and Applications **2000**. Available at: https://www.intechopen.com/books/herbicides-theory-and-applications/the-bioassay-technique-in-the-study-of-the-herbicide-effects

[29]   Pestemer, W.; Stalder, L.; Eckert, B. Availability to plants of herbicide residues in soil. Part II: Data for use in vegetable crop rotations. *Weed Res.,* **1980**, *20*(6), 349-353.
[http://dx.doi.org/10.1111/j.1365-3180.1980.tb00082.x]

[30]   Curi, L.M.; Peltzer, P.M.; Sandoval, M.T.; Lajmanovich, R.C. Acute Toxicity and Sublethal Effects Caused by a Commercial Herbicide Formulated with 2,4-D on Physalaemus albonotatus Tadpoles. *Water Air Soil Pollut.,* **2019**, *230*(1), 22.
[http://dx.doi.org/10.1007/s11270-018-4073-x]

[31]   Anonymous Dicamba Extoxnet. **2014**. Available at: http://pmep.cce.cornell.edu/profiles/extoxnet/carbaryl-dicrotophos/dicamba-ext.html

[32]   Anonymous Imazaquin.Safety Data Sheet. Image 70G Herbicide. **2005**. Available at: http://www.fmccrop.com.au/download/discontinued_FMC/MSDS/dicamba_700_sds_0114.pdf

[33]   Anonymous, Imazethapyr. Toxicology. **2009**. Available at: https://pubchem.ncbi.nlm.nih.gov/compound/Imazethapyr

[34]   Anonymous, Thifensulfuron-Methyl. FAO Specification and Evaluations for Agricultural Pesticides **2011**. Available at:http://www.fao.org/agriculture/crops/core-themes/theme/pests/pm/jmps/ps/en/

[35]   Anonymous, Anonymous. Chlorimuron-ethyl. United States Evironmental Protection Agency Washington **2014**. Available at:https://www3.epa.gov/pesticides/chem_search/ppls/083529-0004--20141002.pdf

[36]   Anonymous, Flumetsulam Safety Data Sheet. DOW Agrosciences LLC. **2015**. Available at:http://www.cdms.net/ldat/mpATA006.pdf

[37]   Anonymous, Sethoxydim Weed Control Methods Handbook, The Nature Conservancy. **2001**. Available at:https://www.invasive.org/gist/products/handbook/methods-handbook.pdf

[38]   Anonymous, Sethoxydim Weed Control Methods Handbook, The Nature Conservancy **2010**. Available at: http://www.fao.org/fileadmin/templates/agphome/documents/Pests_Pesticides/Specs/Clethodim_2017_11_09.pdf

[39]   Anonymous, Trifluralin. European Commission, DG Environment, Brussels **2007**. Available at: https://unece.org/fileadmin/DAM/env/lrtap/TaskForce/popsxg/2008/Trifluralin_RA%20dossier_proposal%20for%20submission%20to%20the%20UNECE%20POP%20Protocol.pdf

[40]   Anonymous, *Fluchloralin solutions.,* **2014**. Available at: https://www.ineos.com/show-document/?grade=Phenol+synthetic++Solutions+with+5+to+15%25+water&bu=INEOS+Phenol&documentT

[41]   Anonymous, Metribuzin. Safety Data Sheet **2016**. Available at: http://www.cdms.net/ldat/mpDH6001.pdf

[42]   Anonymous, Metolachlor. Canadian Water Quality Guidelines for the Protection of Aquatic Life **1999**. Available at: http://ceqg-rcqe.ccme.ca/download/en/193

[43] Ecological Risk Assessment. **2006**. Available at: https://www3.epa.gov/pesticides/chem_ search/cleared_reviews/csr_PC-129051_6-Dec-06_a.pdf

[44] Available from: https://www.who.int/water_sanitation_health/dwq/chemicals/dwq_back-ground_ 20100701_en.pdf

[45] Anonymous, Metribuzin.Safety Data Sheet **2016**. Available at: http://www.cdms.net/ldat/ mpDH6001.pdf

[46] Anonymous, *Linuron DF Herbicide.,* **2016**. Available at: https://agnova.com.au/content/ custom/products/files/Linuron-DF-msds.pdf

[47] Anonymous, Ecological Risk Assessment Problem Formulation For: Fomesafen. 2014. United States Environmental Protection Agency Washington D C. **2014**. Available at: https:// www.epa.gov/sites/production/files/2014-11/documents/eco_risk_assessment1998.pdf

[48] Anonymous, Flumioxazin: Environmental Fate and Ecological Risk Assessment. **2003**. Available at: https://archive.epa.gov/pesticides/chemicalsearch/chemical/foia/web/pdf/129034/129034-2003-- 8-14a.pdf

[49] Available at: https://www3.epa.gov/pesticides/chem_search/reg_actions/reregistrationfs_PC-090501_1-Dec-98.pdf

[50] Anonymous, Metolachlor. Canadian Water Quality Guidelines for the Protection of Aquatic Life. **1999**. Available at: http://ceqg-rcqe.ccme.ca/download/en/193

[51] Anonymous, Chlorpropham. 1996. United States Environmental Protection Agency, R E D. **1996**. Available at: https://archive.epa.gov/pesticides/reregistration/web/pdf/0271red.pdf

[52] Anonymous, Diuron. Environment Assessment. **2011**. Available at: https://apvma. gov.au/sites/default/files/publication/15386-diuron-environment.pdf

[53] Anonymous, Monuron. Safety Data Sheet. **2015**. Available at: http://cdn. chemservice.com/product/msdsnew/External/English/N-12497%20English%20SDS%20US.pdf

[54] Anonymous, Conclusion on the peer review of the pesticide risk assessment of the active substance isoproturon. **2015**. Available at: https://efsa.onlinelibrary.wiley.com/doi/pdf/10.2903/j.efsa.2015.4206

[55] Anonymous, Picloram. FAO Specification and Evaluations for Agricultural Pesticides **2011**. Available at: http://www.fao.org/fileadmin/templates/agphome/documents/Pests_Pesticides/Specs/Picloram_ 2012.pdf

[56] Clopyralid. Safety Data Sheet. 2018. Available at: https://www.caymanchem.com/ msdss/24229m.pdf

[57] Paul, R.; Sharma, R.; Kulshrestha, G.; Singh, S.B. Analysis of metsulfuron-methyl residues in wheat field soil: a comparison of HPLC and bioassay techniques. *Pest Manag. Sci.,* **2009**, *65*(9), 963-968. [http://dx.doi.org/10.1002/ps.1780] [PMID: 19452497]

[58] Diaz, E.C.; Simonet, B.; Valcarcel, M. Liquid-liquid extraction assisted by a carbon nanoparticles interface. Electrophoretic determination of atrazine in environmental samples. *Analyst (Lond.),* **2013**, 1-22.

[59] Sondhia, S. Liquid chromatographic determination of imazosulfuron in food commodity, water and soil using photo diode array detector. *Asian J. Chem.,* **2016**, *28*(5), 1021-1023. [http://dx.doi.org/10.14233/ajchem.2016.19570]

[60] Pirard, C.; Widart, J.; Nguyen, B.K.; Deleuze, C.; Heudt, L.; Haubruge, E.; De Pauw, E.; Focant, J.F. Development and validation of a multi-residue method for pesticide determination in honey using on-column liquid-liquid extraction and liquid chromatography-tandem mass spectrometry. *J. Chromatogr. A,* **2007**, *1152*(1-2), 116-123. [http://dx.doi.org/10.1016/j.chroma.2007.03.035] [PMID: 17416380]

[61] Shamsipur, M.; Fattahi, N.; Pirsaheb, M.; Sharafi, K. Simultaneous preconcentration and determination of 2,4-D, alachlor and atrazine in aqueous samples using dispersive liquid-liquid

microextraction followed by high-performance liquid chromatography ultraviolet detection. *J. Sep. Sci.,* **2012**, *35*(20), 2718-2724.
[http://dx.doi.org/10.1002/jssc.201200424] [PMID: 22997055]

[62]   Wang, C.; Ji, S.; Wu, Q.; Wu, C.; Wang, Z. Determination of triazine herbicides in environmental samples by dispersive liquid-liquid microextraction coupled with high performance liquid chromatography. *J. Chromatogr. Sci.,* **2011**, *49*(9), 689-694.
[http://dx.doi.org/10.1093/chrsci/49.9.689] [PMID: 22586245]

[63]   Moawad, M.; Khoo, C.S. Liquid chromatographic determination of flumetsulam in soybeans. *J. AOAC Int.,* **2005**, *88*(5), 1463-1468.
[http://dx.doi.org/10.1093/jaoac/88.5.1463] [PMID: 16385997]

[64]   Mamun, M.I.R.; Park, J.H.; Choi, J.H.; Kim, H.K.; Choi, W.J.; Han, S.S.; Hwang, K.; Jang, N.I.K.; Assayed, M.E.; El-Dib, M.A.; Shin, H.C.; El-Aty, A.M.A.; Shim, J.H. Development and validation of a multiresidue method for determination of 82 pesticides in water using GC. *J. Sep. Sci.,* **2009**, *32*(4), 559-574.
[http://dx.doi.org/10.1002/jssc.200800606] [PMID: 19212978]

[65]   Mamun, M.I.R.; Park, J.H.; Choi, J.H.; Kim, H.K.; Choi, W.J.; Han, S.S.; Hwang, K.; Jang, N.I.K.; Assayed, M.E.; El-Dib, M.A.; Shin, H.C.; El-Aty, A.M.A.; Shim, J.H. Development and validation of a multiresidue method for determination of 82 pesticides in water using GC. *J. Sep. Sci.,* **2009**, *32*(4), 559-574.
[http://dx.doi.org/10.1002/jssc.200800606] [PMID: 19212978]

[66]   Vidotto, F.; Ferrero, A.; Bertoia, O.; Gennari, M.; Cignetti, A. Dissipation of pretilachlor in paddy water and sediment. *Agronomie,* **2004**, *24*(8), 473-479.
[http://dx.doi.org/10.1051/agro:2004043]

[67]   Mamun, M.I.R.; Park, J.H.; Choi, J.H.; Kim, H.K.; Choi, W.J.; Han, S.S.; Hwang, K.; Jang, N.I.K.; Assayed, M.E.; El-Dib, M.A.; Shin, H.C.; El-Aty, A.M.A.; Shim, J.H. Development and validation of a multiresidue method for determination of 82 pesticides in water using GC. *J. Sep. Sci.,* **2009**, *32*(4), 559-574.
[http://dx.doi.org/10.1002/jssc.200800606] [PMID: 19212978]

[68]   Sondhia, S.; Dixit, A. Bioefficacy and determination of terminal residues of a herbicide anilofos in field soil and plants following an application to the transplanted rice crop. *Commun. Soil Sci. Plant Anal.,* **2015**, *46*(20), 2576-2584.
[http://dx.doi.org/10.1080/00103624.2015.1089261]

[69]   Tandon, S. Degradation kinetics of anilofos in soil and residues in rice crop at harvest. *Pest Manag. Sci.,* **2013**.
[PMID: 24339403]

[70]   Gałuszka, A.; Migaszewski, Z.; Namieśnik, J. The 12 principles of green analytical chemistry and the SIGNIFICANCE mnemonic of green analytical practices. *Trends Analyt. Chem.,* **2013**, *50*, 78-84.
[http://dx.doi.org/10.1016/j.trac.2013.04.010]

[71]   Raynie, D.E. Modern extraction techniques. *Anal. Chem.,* **2010**, *82*(12), 4911-4916.
[http://dx.doi.org/10.1021/ac101223c] [PMID: 20481433]

[72]   Wang, Y.; Sun, Y.; Xu, B.; Li, X.; Jin, R.; Zhang, H.; Song, D. Magnetic ionic liquid-based dispersive liquid-liquid microextraction for the determination of triazine herbicides in vegetable oils by liquid chromatography. *J. Chromatogr. A,* **2014**, *1373*, 9-16.
[http://dx.doi.org/10.1016/j.chroma.2014.11.009] [PMID: 25464995]

[73]   Araujo, L.; Prieto, A.; Troconis, M.; Urribarri, G.; Sandrea, W.; Mercado, J. Determination of acidic herbicides in water samples by in situ derivatization, single drop microextraction and gas chromatography-mass spectrometry. *J. Braz. Chem. Soc.,* **2011**, *22*, 2350-2354.
[http://dx.doi.org/10.1590/S0103-50532011001200016]

[74]   Berrada, H. Application of solid-phase microextraction for determining phenylurea herbicides and

their homologous anilines from vegetables. *J. Chromatography,* **2004**, *1042*(1-2), 9-14.

[75]   Yohannes, A.; Tolesa, T.; Merdassa, Y.; Megersa, N. Single drop microextraction analytical technique for simultaneous separation and trace enrichment of atrazine and its major degradation products from environmental waters followed by liquid chromatographic determination. *J. Anal. Bioanal. Tech.,* **2016**, *7*(5), 5.
[http://dx.doi.org/10.4172/2155-9872.1000330]

[76]   Pano-Farias, N.S.; Ceballos-Magaña, S.G.; Muniz-Valencia, R.; Jurado, J.M.; Alcazar, A. Direct immersion single drop micro-extraction method for multi-class pesticides analysis in mango using GC-MS. *J. AOAC Int.,* **2016**, *82*, 210-219.

[77]   Zhao, E.C.; Shan, W.L.; Jiang, S.R.; Liu, Y.; Zhou, Z.Q. Determination of the chloroacetanilide herbicides in waters using single-drop microextraction and gas chromatography. *Chem. Pap.,* **2006**, *83*, 105-110.

[78]   Tsiropoulos, N.G.; Amvrazi, E.G. Determination of pesticide residues in honey by single-drop microextraction and gas chromatography. *J. AOAC Int.,* **2011**, *94*(2), 634-644.
[http://dx.doi.org/10.1093/jaoac/94.2.634] [PMID: 21563700]

[79]   Hu, M.; Chen, H.; Jiang, Y.; Zhu, H. Headspace single-drop microextraction coupled with gas chromatography electron capture detection of butanone derivative for determination of iodine in milk powder and urine. *Chem. Pap.,* **2013**, *67*(10), 1255-1261.
[http://dx.doi.org/10.2478/s11696-013-0391-z]

[80]   Wielgomas, B.; Czarnowski, W. Headspace single-drop microextraction and GC-ECD determination of chlorpyrifos-ethyl in rat liver. *Anal. Bioanal. Chem.,* **2008**, *390*(7), 1933-1941.
[http://dx.doi.org/10.1007/s00216-008-1831-4] [PMID: 18299820]

[81]   He, Y.; Lee, H.K. Continuous flow microextraction combined with high-performance liquid chromatography for the analysis of pesticides in natural waters. *J. Chromatogr. A,* **2006**, *1122*(1-2), 7-12.
[http://dx.doi.org/10.1016/j.chroma.2006.04.078] [PMID: 16716335]

[82]   Chen, D.; Zhang, Y.; Miao, H.; Zhao, Y.; Wu, Y. Determination of triazine herbicides in drinking water by dispersive micro solid phase extraction with ultrahigh-performance liquid chromatography-high-resolution mass spectrometric detection. *J. Agric. Food Chem.,* **2015**, *63*(44), 9855-9862.
[http://dx.doi.org/10.1021/acs.jafc.5b03973] [PMID: 26487365]

[83]   Bol'shakov, D.S.; Amelin, V.G.; Tret'yakov, A.V. Determination of herbicides and their metabolites in natural waters by capillary zone electrophoresis combined with dispersive liquid-liquid microextraction and on-line preconcentration. *J. Anal. Chem.,* **2014**, *69*(1), 72-82.
[http://dx.doi.org/10.1134/S106193481311004X]

[84]   Farajzadeh, M.A.; Mohebbi, A.; Feriduni, B. Development of continuous dispersive liquid-liquid microextraction performed in home-made device for extraction and preconcentration of aryloxyphenoxy-propionate herbicides from aqueous samples followed by gas chromatography-flame ionization detection. *Anal. Chim. Acta,* **2016**, *920*, 1-9.
[http://dx.doi.org/10.1016/j.aca.2016.03.041] [PMID: 27114217]

[85]   Megersa, N. Hollow fiber-liquid phase microextraction for trace enrichment of the residues of atrazine and its major degradation products from environmental water and human urine samples. *Anal. Methods,* **2015**, *7*(23), 9940-9948.
[http://dx.doi.org/10.1039/C5AY01927C]

[86]   Asensio-Ramos, M.; Hernández-Borges, J.; González-Hernández, G.; Rodríguez-Delgado, M.Á. Hollow-fiber liquid-phase microextraction for the determination of pesticides and metabolites in soils and water samples using HPLC and fluorescence detection. *Electrophoresis,* **2012**, *33*(14), 2184-2191.
[http://dx.doi.org/10.1002/elps.201200138] [PMID: 22821496]

[87]   Xu, H.; Pan, W.; Song, D.; Yang, G. Development of an improved liquid phase microextraction technique and its application in the analysis of flumetsulam and its two analogous herbicides in soil. *J.*

Agric. Food Chem., **2007**, *55*(23), 9351-9356.
[http://dx.doi.org/10.1021/jf0718345] [PMID: 17953444]

[88]   Gure, A.; Lara, F.J.; Moreno-González, D.; Megersa, N.; del Olmo-Iruela, M.; García-Campaña, A.M. Salting-out assisted liquid-liquid extraction combined with capillary HPLC for the determination of sulfonylurea herbicides in environmental water and banana juice samples. *Talanta*, **2014**, *127*, 51-58.
[http://dx.doi.org/10.1016/j.talanta.2014.03.070] [PMID: 24913856]

[89]   Teju, E.; Tadesse, B.; Megersa, N. Salting-out assisted liquid-liquid extraction for the preconcentration and quantitative determination of 8 herbicide residues simultaneously in different water samples with high performance liquid chromatography. *Sep Sci Technol. J Food Measure Character*, **2017**, pp. 719-729.

[90]   Rashidipour, M.; Heydari, R.; Maleki, A.; Mohammadi, E.; Davari, B. Salt-assisted liquid-liquid extraction coupled with reversed-phase dispersive liquid-liquid microextraction for sensitive HPLC determination of paraquat in environmental and food samples. *J. Food Meas. Charact.*, **2019**, *13*(1), 269-276.
[http://dx.doi.org/10.1007/s11694-018-9941-y]

[91]   Asensio-Ramos, M.; Hernández-Borges, J.; González-Hernández, G.; Rodríguez-Delgado, M.Á. Hollow-fiber liquid-phase microextraction for the determination of pesticides and metabolites in soils and water samples using HPLC and fluorescence detection. *Electrophoresis*, **2012**, *33*(14), 2184-2191.
[http://dx.doi.org/10.1002/elps.201200138] [PMID: 22821496]

[92]   Gure, A.; Lara, F.J.; Megersa, N.; del Olmo-Iruela, M.; García-Campaña, A.M.; Garcia-Campana, A.M. Dispersive liquid-liquid microextraction followed by capillary high-performance liquid chromatography for the determination of six sulfonylurea herbicides in fruit juices. *Food Anal. Methods*, **2013**.
[http://dx.doi.org/10.1007/s12161-013-9775-5]

[93]   Suarez, L.A.; Apan, A.; Werth, J. Detection of phenoxy herbicide dosage in cotton crops through the analysis of hyperspectral data. *Int. J. Remote Sens.*, **2017**, *38*(23), 6528-6553.
[http://dx.doi.org/10.1080/01431161.2017.1362128]

[94]   Gure, A; Lara, F J; Moreno-González , D; Megersa, N; Olmo-Iruela, M D; Garcia-Campana, A.M. Salting-out assisted liquid-liquid extraction combined with capillary HPLC for the determination of sulfonylurea herbicides in environmental water and banana juice samples. *Talanta*, **2014**, 51-58.

[95]   Teju, E.; Tadesse, B.; Megersa, N. Salting-out assisted liquid-liquid extraction for the preconcentration and quantitative determination of 8 herbicide residues simultaneously in different water samples with high performance liquid chromatography. *Sep. Sci. Technol.*, **2017**, 719-729.

[97]   Razmara, R.S.; Daneshfar, A.; Sahrai, R. Determination of methylene blue and sunset yellow in wastewater and food samples using salting-out assisted liquid-liquid extraction. *J. Ind. Eng. Chem.*, **2011**, *17*(3), 533-536.
[http://dx.doi.org/10.1016/j.jiec.2010.10.028]

[97]   Kumari, A.; Vuppala, S.; Satyavathi, B. Salting-Out Assisted Liquid-Liquid Extraction (SALLE) for the separation of morpholine from aqueous stream: Phase equilibrium, optimization and modeling. *J. Mol. Liq.*, **2019**, *111*, 884.

[98]   Jain, D.; Athawale, R.; Bajaj, A.; Shrikhande, S. Double-salting out assisted liquid-liquid extraction (SALLE) HPLC method for estimation of temozolomide from biological samples. *J. Chromatogr. B Analyt. Technol. Biomed. Life Sci.*, **2014**, *970*, 86-94.
[http://dx.doi.org/10.1016/j.jchromb.2014.02.031] [PMID: 25240926]

[99]   Jabusch, T.W.; Tjeerdema, R.S. Microbial degradation of penoxsulam in flooded rice field soils. *J. Agric. Food Chem.*, **2006**, *54*(16), 5962-5967.
[http://dx.doi.org/10.1021/jf0606454] [PMID: 16881702]

[100] Vargas-Pérez, M.; González, F.J.E.; Frenich, A.G. Evaluation of the behaviour of propoxycarbazone herbicide in soils and water under different conditions. Post-targeted study. *Ecotoxicol. Environ. Saf.*,

2019, *183*, 109506.
[http://dx.doi.org/10.1016/j.ecoenv.2019.109506] [PMID: 31386940]

[101] Carabias-Martínez, R.; Rodríguez-Gonzalo, E.; Herrero-Hernández, E.; Sánchez-San Román, F.J.; Prado Flores, M.G. Determination of herbicides and metabolites by solid-phase extraction and liquid chromatography. *J. Chromatogr. A*, **2002**, *950*(1-2), 157-166.
[http://dx.doi.org/10.1016/S0021-9673(01)01613-2] [PMID: 11990989]

[102] Rodríguez-González, N.; Beceiro-González, E.; González-Castro, M.J.; Alpendurada, M.F. On-line solid-phase extraction method for determination of triazine herbicides and degradation products in seawater by ultra-pressure liquid chromatography-tandem mass spectrometry. *J. Chromatogr. A*, **2016**, *1470*, 33-41.
[http://dx.doi.org/10.1016/j.chroma.2016.10.007] [PMID: 27726863]

[103] Su, Y.; Wang, S.; Zhang, N.; Cui, P.; Gao, Y.; Bao, T. Zr-MOF modified cotton fiber for pipette tip solid-phase extraction of four phenoxy herbicides in complex samples. *Ecotoxicol. Environ. Saf.*, **2020**, *201*, 110764.
[http://dx.doi.org/10.1016/j.ecoenv.2020.110764] [PMID: 32480162]

[104] Wang, J.; Leung, D.; Chow, W.; Wong, J.W.; Chang, J. UHPLC/ESI Q-Orbitrap Quantitation of 655 Pesticide Residues in Fruits and Vegetables - a Companion to a DATA Working Flow. *J of AOAC Interna*, **2020**, *5*, 68-75.

[105] Qin, Z.; Jiang, Y.; Piao, H.; Li, J.; Tao, S.; Ma, P.; Wang, X.; Song, D.; Sun, Y. MIL-101(Cr)/MWCNTs-functionalized melamine sponges for solid-phase extraction of triazines from corn samples, and their subsequent determination by HPLC-MS/MS. *Talanta*, **2020**, *211*, 120676.
[http://dx.doi.org/10.1016/j.talanta.2019.120676] [PMID: 32070600]

[106] Zheng, X.; Wang, J. A Novel Metal-Organic Framework Composite, MIL-101(Cr)@MIP, as an Efficient Sorbent in Solid-Phase Extraction Coupling with HPLC for Tribenuron-Methyl Determination. *Intern J Anal Chem*, **2019**, 1-10.

[107] Guo, L.; Deng, Q.; Fang, G.; Gao, W.; Wang, S. Preparation and evaluation of molecularly imprinted ionic liquids polymer as sorbent for on-line solid-phase extraction of chlorsulfuron in environmental water samples. *J. Chromatogr. A*, **2011**, *1218*(37), 6271-6277.
[http://dx.doi.org/10.1016/j.chroma.2011.07.016] [PMID: 21807367]

[108] Barchanska, H.; Rusek, M.; Szatkowska, A. New procedures for simultaneous determination of mesotrione and atrazine in water and soil. Comparison of the degradation processes of mesotrione and atrazine. *Environ. Monit. Assess.*, **2012**, *184*(1), 321-334.
[http://dx.doi.org/10.1007/s10661-011-1970-5] [PMID: 21416215]

[109] Shah, J.; Jan, M.R.; Muhammad, M.; Shehzad, F.N. Development of a complex-based flow injection spectrophotometric method for determination of the herbicide pinoxaden in environmental samples. *Toxicol. Environ. Chem.*, **2011**, *93*(8), 1547-1556.
[http://dx.doi.org/10.1080/02772248.2011.604323]

[110] Apprao, K.; Babu, S. M.S, Rao B, M.V and N Rao N, T. A new analytical approach for the determination of triafamone and its metabolites residue in wheat plant. *Int. J. Curr. Adv. Res.*, **2016**, *8*, 38393-38397.

[111] Biesaga, M.; Jankowska, A.; Pyrzyńska, K. Comparison of different sorbents for solid-phase extraction of phenoxyalkanoic acid herbicides. *Mikrochim. Acta*, **2005**, *150*(3-4), 317-322.
[http://dx.doi.org/10.1007/s00604-005-0359-y]

[112] Fenoll, J.; Hellín, P.; Martínez, C.M.; Flores, P.; Navarro, S. Semiconductor-sensitized photodegradation of s-triazine and chloroacetanilide herbicides in leaching water using TiO_2 and ZnO as catalyst under natural sunlight. *J. Photochem. Photobiol. Chem.*, **2012**, *238*, 81-87.
[http://dx.doi.org/10.1016/j.jphotochem.2012.04.017]

[113] Mizajani, R.; Ramezani, Z.; Kardan, F. Selective determination of thidiazuron herbicide in fruit and vegetable samples using molecularly imprinted polymer fiber solid phase microextraction with ion

mobility spectrometry detection (MIPF-SPME-IMS). *Microchem. J.,* **2016**, *16*, 30301-0.

[114] Zhou, Q.; Wang, W.; Xiao, J. Preconcentration and determination of nicosulfuron, thifensulfuron-methyl and metsulfuron-methyl in water samples using carbon nanotubes packed cartridge in combination with high performance liquid chromatography. *Anal. Chim. Acta,* **2006**, *559*(2), 200-206.
[http://dx.doi.org/10.1016/j.aca.2005.11.079]

[115] Min, G.; Wang, S.; Zhu, H.; Fang, G.; Zhang, Y. Multi-walled carbon nanotubes as solid-phase extraction adsorbents for determination of atrazine and its principal metabolites in water and soil samples by gas chromatography-mass spectrometry. *Sci. Total Environ.,* **2008**, *396*(1), 79-85.
[http://dx.doi.org/10.1016/j.scitotenv.2008.02.016] [PMID: 18372005]

[116] Valimaña-Traverso, J.; Morante-Zarcero, S.; Pérez-Quintanilla, D.; García, M.Á.; Sierra, I.; Marina, M.L. Cationic amine-bridged periodic mesoporous organosilica materials for off-line solid-phase extraction of phenoxy acid herbicides from water samples prior to their simultaneous enantiomeric determination by capillary electrophoresis. *J. Chromatogr. A,* **2018**, *1566*, 146-157.
[http://dx.doi.org/10.1016/j.chroma.2018.06.042] [PMID: 29960738]

[117] Lee, J.H.; Park, H.N.; Park, H.J.; Heo, S.; Park, S.S.; Park, S.K.; Baek, S.Y. Development and Validation of LC-MS/MS and LC-Q-Orbitrap/MS Methods for Determination of Glyphosate in Vaccines. *Chromatographia,* **2017**, *80*(12), 1741-1747.
[http://dx.doi.org/10.1007/s10337-017-3417-9]

[118] Spietelun, A.; Kloskowski, A.; Chrzanowski, W.; Namieśnik, J. Understanding solid-phase microextraction: key factors influencing the extraction process and trends in improving the technique. *Chem. Rev.,* **2013**, *113*(3), 1667-1685.
[http://dx.doi.org/10.1021/cr300148j] [PMID: 23273266]

[119] Silva, F.A.; Chawla, N.; Filho, R.D.T. Tensile behavior of high performance natural (sisal) fibers. *Compos. Sci. Technol.,* **2008**, *68*(15-16), 3438-3443.
[http://dx.doi.org/10.1016/j.compscitech.2008.10.001]

[120] Kataoka, H.; Mitani, K.; Takino, M. Analysis of herbicides in water by on-line in-tube solid-phase microextraction coupled with liquid chromatography-mass spectrometry. *Pesticide Protocols.,* **2006**, *19*, 365-382.
[http://dx.doi.org/10.1385/1-59259-929-X:365]

[121] Neng, N.R.; Mestre, A.S.; Carvalho, A.P.; Nogueira, J.M.F. Powdered activated carbons as effective phases for bar adsorptive micro-extraction (BAμE) to monitor levels of triazinic herbicides in environmental water matrices. *Talanta,* **2011**, *83*(5), 1643-1649.
[http://dx.doi.org/10.1016/j.talanta.2010.11.049] [PMID: 21238763]

[122] Assoumani, A.; Lissalde, S.; Margoum, C.; Mazzella, N.; Coquery, M. *In situ* application of stir bar sorptive extraction as a passive sampling technique for the monitoring of agricultural pesticides in surface waters. *Sci. Total Environ.,* **2013**, *463-464*, 829-835.
[http://dx.doi.org/10.1016/j.scitotenv.2013.06.025] [PMID: 23856404]

[123] Kawaguchi, M.; Ishii, Y.; Sakui, N.; Okanouchi, N.; Ito, R.; Saito, K.; Nakazawa, H. Stir bar sorptive extraction with *in situ* derivatization and thermal desorption-gas chromatography-mass spectrometry for determination of chlorophenols in water and body fluid samples. *Anal. Chim. Acta,* **2005**, *533*(1), 57-65.
[http://dx.doi.org/10.1016/j.aca.2004.10.080]

[124] Kiljanek, T.; Niewiadowska, A.; Semeniuk, S.; Gaweł, M.; Borzęcka, M.; Posyniak, A. Multi-residue method for the determination of pesticides and pesticide metabolites in honeybees by liquid and gas chromatography coupled with tandem mass spectrometry—Honeybee poisoning incidents. *J. Chromatogr. A,* **2016**, *1435*, 100-114.
[http://dx.doi.org/10.1016/j.chroma.2016.01.045] [PMID: 26830634]

[125] Bouri, M.; Gurau, M.; Salghi, R.; Cretescu, I.; Zougagh, M.; Rios, Á. Ionic liquids supported on magnetic nanoparticles as a sorbent preconcentration material for sulfonylurea herbicides prior to their

determination by capillary liquid chromatography. *Anal. Bioanal. Chem.,* **2012**, *404*(5), 1529-1538.
[http://dx.doi.org/10.1007/s00216-012-6221-2] [PMID: 22832671]

[126] Zheng, S.; He, M.; Chen, B.; Hu, B. Porous aromatic framework coated stir bar sorptive extraction coupled with high performance liquid chromatography for the analysis of triazine herbicides in maize samples. *J. Chromatogr. A,* **2020**, *1614*, 460728.
[http://dx.doi.org/10.1016/j.chroma.2019.460728] [PMID: 31785896]

[127] Assoumani, A.; Lissalde, S.; Margoum, C.; Mazzella, N.; Coquery, M. *In situ* application of stir bar sorptive extraction as a passive sampling technique for the monitoring of agricultural pesticides in surface waters. *Sci. Total Environ.,* **2013**, *463-464*, 829-835.
[http://dx.doi.org/10.1016/j.scitotenv.2013.06.025] [PMID: 23856404]

[128] Sanchez-Ortega, A.; Unceta, N.; Gómez-Caballero, A.; Sampedro, M.C.; Akesolo, U.; Goicolea, M.A.; Barrio, R.J. Sensitive determination of triazines in underground waters using stir bar sorptive extraction directly coupled to automated thermal desorption and gas chromatography-mass spectrometry. *Anal. Chim. Acta,* **2009**, *641*(1-2), 110-116.
[http://dx.doi.org/10.1016/j.aca.2009.03.044] [PMID: 19393374]

[129] Liang, L.; Wang, X.; Sun, Y.; Ma, P.; Li, X.; Piao, H.; Jiang, Y.; Song, D. Magnetic solid-phase extraction of triazine herbicides from rice using metal-organic framework MIL-101(Cr) functionalized magnetic particles. *Talanta,* **2018**, *179*, 512-519.
[http://dx.doi.org/10.1016/j.talanta.2017.11.017] [PMID: 29310269]

[130] Ma, J.; Jiang, L.; Wu, G.; Xia, Y.; Lu, W.; Li, J.; Chen, L. Determination of six sulfonylurea herbicides in environmental water samples by magnetic solid-phase extraction using multi-walled carbon nanotubes as adsorbents coupled with high-performance liquid chromatography. *J. Chromatogr. A,* **2016**, *1466*, 12-20.
[http://dx.doi.org/10.1016/j.chroma.2016.08.065] [PMID: 27590086]

[131] Wang, S.Y.; Kong, C.; Chen, Q.P.; Yu, H.J. Screening 89 pesticides in fishery drugs by ultrahigh performance liquid chromatography tandem quadrupole-orbitrap mass spectrometer. *Molecules,* **2019**, *24*(18), 3375.
[http://dx.doi.org/10.3390/molecules24183375] [PMID: 31533222]

[132] He, Z.; Liu, D.; Li, R.; Zhou, Z.; Wang, P. Magnetic solid-phase extraction of sulfonylurea herbicides in environmental water samples by Fe_3O_4@dioctadecyl dimethyl ammonium chloride@silica magnetic particles. *Anal. Chim. Acta,* **2012**, *747*, 29-35.
[http://dx.doi.org/10.1016/j.aca.2012.08.015] [PMID: 22986132]

[133] Li, N.; Zhang, L.; Nian, L.; Cao, B.; Wang, Z.; Lei, L.; Yang, X.; Sui, J.; Zhang, H.; Yu, A. Dispersive micro-solid-phase extraction of herbicides in vegetable oil with metal-organic framework MIL-101. *J. Agric. Food Chem.,* **2015**, *63*(8), 2154-2161.
[http://dx.doi.org/10.1021/jf505760y] [PMID: 25665636]

[134] Chen, D.; Zhang, Y.; Miao, H.; Zhao, Y.; Wu, Y. Determination of triazine herbicides in drinking water by dispersive micro solid phase extraction with ultrahigh-performance liquid chromatography-high-resolution mass spectrometric detection. *J. Agric. Food Chem.,* **2015**, *63*(44), 9855-9862.
[http://dx.doi.org/10.1021/acs.jafc.5b03973] [PMID: 26487365]

[135] Zhang, Y.; Chen, D.; Zhao, Y. Determination of triazine pesticides in honey by ultra high performance liquid chromatography-high resolution isotope dilution mass spectrometry combined with dispersive micro-solid phase extraction. *Anal. Methods,* **2015**, *7*(23), 9867-9874.
[http://dx.doi.org/10.1039/C5AY01366F]

[136] Luo, M.; Liu, D.; Zhao, L.; Han, J.; Liang, Y.; Wang, P.; Zhou, Z. A novel magnetic ionic liquid modified carbon nanotube for the simultaneous determination of aryloxyphenoxy-propionate herbicides and their metabolites in water. *Anal. Chim. Acta,* **2014**, *852*, 88-96.
[http://dx.doi.org/10.1016/j.aca.2014.09.024] [PMID: 25441884]

[137] Zhang, G; Zang, X; Li, Z; Chang, Q; Wang, C; Wang, Z Solid phase microextraction using a

grapheme composite-coated fiber coupled with gas chromatography for the determination of acetanilide herbicides in water samples **2014**, *6*, 2756.

[138]   Wang, X.; Li, X.; Li, Z.; Zhang, Y.; Bai, Y.; Liu, H. Online coupling of in-tube solid-phase microextraction with direct analysis in real time mass spectrometry for rapid determination of triazine herbicides in water using carbon-nanotubes-incorporated polymer monolith. *Anal. Chem.,* **2014**, *86*(10), 4739-4747.
[http://dx.doi.org/10.1021/ac500382x] [PMID: 24745793]

[139]   Rocha, C.; Pappas, E.A.; Huang, C. Determination of trace triazine and chloroacetamide herbicides in tile-fed drainage ditch water using solid-phase microextraction coupled with GC-MS. *Environ. Pollut.,* **2008**, *152*(1), 239-244.
[http://dx.doi.org/10.1016/j.envpol.2007.04.029] [PMID: 17629381]

[140]   Wu, Q.; Feng, C.; Zhao, G.; Wang, C.; Wang, Z. Graphene-coated fiber for solid-phase microextraction of triazine herbicides in water samples. *J. Sep. Sci.,* **2012**, *35*(2), 193-199.
[http://dx.doi.org/10.1002/jssc.201100740] [PMID: 22162195]

[141]   Mirzajani, R.; Ramezani, Z.; Kardani, F. Selective determination of thidiazuron herbicide in fruit and vegetable samples using molecularly imprinted polymer fiber solid phase microextraction with ion mobility spectrometry detection (MIPF-SPME-IMS). *Microchem. J.,* **2017**, *130*, 93-101.
[http://dx.doi.org/10.1016/j.microc.2016.08.009]

[142]   Barker, S.A. Matrix solid phase dispersion (MSPD). *J. Biochem. Biophys. Methods,* **2007**, *70*(2), 151-162.
[http://dx.doi.org/10.1016/j.jbbm.2006.06.005] [PMID: 17107714]

[143]   Capriotti, A.L.; Cavaliere, C.; Giansanti, P.; Gubbiotti, R.; Samperi, R.; Laganà, A. Recent developments in matrix solid-phase dispersion extraction. *J. Chromatogr. A,* **2010**, *1217*(16), 2521-2532.
[http://dx.doi.org/10.1016/j.chroma.2010.01.030] [PMID: 20116066]

[144]   Kishida, K.; Furusawa, N. Matrix solid-phase dispersion extraction and high-performance liquid chromatographic determination of residual sulfonamides in chicken. *J. Chromatogr. A,* **2001**, *937*(1-2), 49-55.
[http://dx.doi.org/10.1016/S0021-9673(01)01307-3] [PMID: 11765084]

[145]   Capriotti, A.L.; Cavaliere, C.; Giansanti, P.; Gubbiotti, R.; Samperi, R.; Laganà, A. Recent developments in matrix solid-phase dispersion extraction. *J. Chromatogr. A,* **2010**, *1217*(16), 2521-2532.
[http://dx.doi.org/10.1016/j.chroma.2010.01.030] [PMID: 20116066]

[146]   Cai, C.; Cheng, H.; Wang, Y. Determination of pretilachlor in soil and rice using matrix solid-phase dispersion extraction by capillary electrophoresis with field amplified sample injection and electrochemiluminescence detection. *Anal. Methods,* **2014**, *6*(8), 2767-2773.
[http://dx.doi.org/10.1039/c3ay41950a]

[147]   Kalsi, N.K.; Kaur, P. Dissipation of bispyribac sodium in aridisols: Impact of soil type, moisture and temperature. *Ecotoxicol. Environ. Saf.,* **2019**, *170*, 375-382.
[http://dx.doi.org/10.1016/j.ecoenv.2018.12.005] [PMID: 30550967]

[148]   Barchanska, H.; Rusek, M.; Szatkowska, A. New procedures for simultaneous determination of mesotrione and atrazine in water and soil. Comparison of the degradation processes of mesotrione and atrazine. *Environ. Monit. Assess.,* **2012**, *184*(1), 321-334.
[http://dx.doi.org/10.1007/s10661-011-1970-5] [PMID: 21416215]

[149]   Piao, H.; Jiang, Y.; Li, X.; Ma, P.; Wang, X.; Song, D.; Sun, Y. Matrix solid-phase dispersion coupled with hollow fiber liquid phase microextraction for determination of triazine herbicides in peanuts. *J. Sep. Sci.,* **2019**, *42*(12), 2123-2130.
[http://dx.doi.org/10.1002/jssc.201801213] [PMID: 30980589]

[150]   Liang, L.; Wang, X.; Sun, Y.; Ma, P.; Li, X.; Piao, H.; Jiang, Y.; Song, D. Magnetic solid-phase

extraction of triazine herbicides from rice using metal-organic framework MIL-101(Cr) functionalized magnetic particles. *Talanta,* **2018**, *179*, 512-519.
[http://dx.doi.org/10.1016/j.talanta.2017.11.017] [PMID: 29310269]

[151] Barchanska, H.; Rusek, M.; Szatkowska, A. New procedures for simultaneous determination of mesotrione and atrazine in water and soil. Comparison of the degradation processes of mesotrione and atrazine. *Environ. Monit. Assess.,* **2012**, *184*(1), 321-334.
[http://dx.doi.org/10.1007/s10661-011-1970-5] [PMID: 21416215]

[152] Tsochatzis, E.D.; Menkissoglu-Spiroudi, U.; Karpouzas, D.G.; Tzimou-Tsitouridou, R. A multi-residue method for pesticide residue analysis in rice grains using matrix solid-phase dispersion extraction and high-performance liquid chromatography-diode array detection. *Anal. Bioanal. Chem.,* **2010**, *397*(6), 2181-2190.
[http://dx.doi.org/10.1007/s00216-010-3645-4] [PMID: 20379813]

[153] Kaur, P.; Kaur, H.; Bhullar, M.S. Influence of organic amendments on dissipation and leaching potential of penoxsulam under laboratory conditions. Int J Environ Anal Chem. *NAAS,* **2020**, *7*, 43.
[http://dx.doi.org/10.1080/03067319.2020.1748611]

[154] Makkar, A.; Kaur, P.; Kaur, P.; Kaur, K. Comparison of extraction techniques for quantitative analysis of pendimethalin from soil and rice grain. *J. Liq. Chromatogr. Relat. Technol.,* **2016**, *39*(15), 718-723.
[http://dx.doi.org/10.1080/10826076.2016.1238392]

[155] Rojano-Delgado, A.M.; Priego-Capote, F.; Prado, R.D.; de Castro, M.D.L. Ultrasound-assisted extraction with LC-TOF/MS identification and LC-UV determination of imazamox and its metabolites in leaves of wheat plants. *Phytochem. Anal.,* **2014**, *25*(4), 357-363.
[http://dx.doi.org/10.1002/pca.2467] [PMID: 23934624]

[156] Shehzad, F.N.; Shah, J. Quantification of pinoxaden herbicide in wheat grains and vegetable samples by ultrasonication-assisted extraction and high-performance liquid chromatography. *Pak. J. Weed Sci. Res.,* **2013**, *19*, 167-177.

[157] Huang, Z.; Meng, X.; Liu, M.; Wang, S. A hydroxyl functionalized ionic liquid-based ultrasound-assisted surfactant-enhanced emulsification microextraction for the determination of herbicides in water samples. *Anal. Methods,* **2014**, *6*(21), 8744-8751.
[http://dx.doi.org/10.1039/C4AY01198H]

[158] Ghobadi, M.; Yamini, Y.; Ebrahimpour, B. Extraction and determination of sulfonylurea herbicides in water and soil samples by using ultrasound-assisted surfactant-enhanced emulsification microextraction and analysis by high-performance liquid chromatography. *Ecotoxicol. Environ. Saf.,* **2015**, *112*, 68-73.
[http://dx.doi.org/10.1016/j.ecoenv.2014.09.035] [PMID: 25463855]

[159] Seebunrueng, K.; Santaladchaiyakit, Y.; Srijaranai, S. The simultaneous analysis of sulfonylurea herbicide residues in fruit samples using ultrasound-assisted surfactant-enhanced emulsification microextraction coupled with high-performance liquid chromatography. *Anal. Methods,* **2013**, *5*(21), 6009-6016.
[http://dx.doi.org/10.1039/c3ay40096d]

[160] Yan, R.; Ju, F.; Wang, H.; Sun, C.; Zhang, H.; Shao, M.; Wang, Y. Determination of sulfonylurea herbicides in soil by ionic liquid-based ultrasonic-assisted extraction high-performance liquid chromatography. *Anal. Methods,* **2014**, *6*(24), 9561-9566.
[http://dx.doi.org/10.1039/C4AY01876A]

[161] Yan, R.; Ju, F.; Wang, H.; Sun, C.; Zhang, H.; Shao, M.; Wang, Y. Determination of sulfonylurea herbicides in soil by ionic liquid-based ultrasonic-assisted extraction high-performance liquid chromatography. *Anal. Methods,* **2014**, *6*(24), 9561-9566.
[http://dx.doi.org/10.1039/C4AY01876A]

[162] Asiabi, H.; Yamini, Y.; Moradi, M. Determination of sulfonylurea herbicides in soil samples *via* supercritical fluid extraction followed by nanostructured supramolecular solvent microextraction. *J.*

Supercrit. Fluids, **2013**, *84*, 20-28.
[http://dx.doi.org/10.1016/j.supflu.2013.09.008]

[163]   Gonçalves, C.; Carvalho, J.J.; Azenha, M.A.; Alpendurada, M.F. Optimization of supercritical fluid extraction of pesticide residues in soil by means of central composite design and analysis by gas chromatography-tandem mass spectrometry. *J. Chromatogr. A,* **2006**, *1110*(1-2), 6-14.
[http://dx.doi.org/10.1016/j.chroma.2006.01.089] [PMID: 16480994]

[164]   Sun, L.; Lee, H.K. Optimization of microwave-assisted extraction and supercritical fluid extraction of carbamate pesticides in soil by experimental design methodology. *J. Chromatogr. A,* **2003**, *1014*(1-2), 165-177.
[http://dx.doi.org/10.1016/S0021-9673(03)00574-0] [PMID: 14558622]

[165]   Rissato, S.; Galhiane, M.; Knoll, F.; Apon, B. Supercritical fluid extraction for pesticide multiresidue analysis in honey: determination by gas chromatography with electron-capture and mass spectrometry detection. *J. Chromatogr. A,* **2004**, *1048*(2), 153-159.
[http://dx.doi.org/10.1016/S0021-9673(04)01213-0] [PMID: 15481252]

[166]   Kang, C.A.; Kim, M.R.; Shen, J.Y.; Cho, I.K.; Park, B.J.; Kim, I.S.; Shim, J.H. Supercritical fluid extraction for liquid chromatographic determination of pyrazosulfuron-ethyl in soils. *Bull. Environ. Contam. Toxicol.,* **2006**, *76*(5), 745-751.
[http://dx.doi.org/10.1007/s00128-006-0983-1] [PMID: 16786443]

[167]   DeMartinis, B.S.; Lanças, F.M. An alternative supercritical fluid extraction system for aqueous matrices and its application in pesticides residue analysis. *J. Environ. Sci. Health B,* **2000**, *35*(5), 539-547.
[http://dx.doi.org/10.1080/03601230009373290] [PMID: 10968605]

[168]   Senseman, S.A.; Ketchersid, M.L. Evaluation of Co-solvents with supercritical fluid extraction of atrazine from soil. *Arch. Environ. Contam. Toxicol.,* **2000**, *38*(3), 263-267.
[http://dx.doi.org/10.1007/s002449910035] [PMID: 10667922]

[169]   Taylor, P.N.; Scrimshaw, M.D.; Lester, J.N. Supercitical fluid extraction of acidic herbicides from sediment. *Inter Envimn Awl Chem,* **1996**, *69*, 141-155.

[170]   Tolcha, T.; Gemechu, T.; Al-Hamimi, S.; Megersa, N.; Turner, C. High density supercritical carbon dioxide for the extraction of pesticide residues in onion with multivariate response surface methodology. *Molecules,* **2020**, *25*(4), 1012.
[http://dx.doi.org/10.3390/molecules25041012] [PMID: 32102410]

[171]   Berglof, T. SFE Supercritical Fluid Extraction in Environmental Analysis: Total extraction and study of pesticide retention in soil. **1997**.

[172]   Asiabi, H.; Yamini, Y.; Moradi, M. Determination of sulfonylurea herbicides in soil samples *via* supercritical fluid extraction followed by nanostructured supramolecular solvent microextraction. *J. Supercrit. Fluids,* **2013**, *84*, 20-28.
[http://dx.doi.org/10.1016/j.supflu.2013.09.008]

[173]   Zhou, T.; Ding, J.; Wang, Q.; Xu, Y.; Wang, B.; Zhao, L.; Ding, H.; Chen, Y.; Ding, L. Microwave-assisted rapid preparation of monodisperse superhydrophilic resin microspheres as adsorbent for triazines in fruit juices. *Talanta,* **2018**, *179*, 734-741.
[http://dx.doi.org/10.1016/j.talanta.2017.12.003] [PMID: 29310301]

[174]   Wu, L.; Hu, M.; Li, Z.; Song, Y.; Yu, C.; Zhang, Y.; Zhang, H.; Yu, A.; Ma, Q.; Wang, Z. Determination of triazine herbicides in fresh vegetables by dynamic microwave-assisted extraction coupled with homogeneous ionic liquid microextraction high performance liquid chromatography. *Anal. Bioanal. Chem.,* **2015**, *407*(6), 1753-1762.
[http://dx.doi.org/10.1007/s00216-014-8393-4] [PMID: 25542578]

[175]   Li, N.; Wu, L.; Nian, L.; Song, Y.; Lei, L.; Yang, X.; Wang, K.; Wang, Z.; Zhang, L.; Zhang, H.; Yu, A.; Zhang, Z. Dynamic microwave assisted extraction coupled with dispersive micro-solid-phase extraction of herbicides in soybeans. *Talanta,* **2015**, *142*, 43-50.

[http://dx.doi.org/10.1016/j.talanta.2015.04.038] [PMID: 26003690]

[176] Wang, H.; Li, G.; Zhang, Y.; Chen, H.; Zhao, Q.; Song, W.; Xu, Y.; Jin, H.; Ding, L. Determination of triazine herbicides in cereals using dynamic microwave-assisted extraction with solidification of floating organic drop followed by high-performance liquid chromatography. *J. Chromatogr. A,* **2012,** *1233,* 36-43.
[http://dx.doi.org/10.1016/j.chroma.2012.02.034] [PMID: 22402128]

[177] Barchanska, H.; Sajdak, M.; Szczypka, K.; Swientek, A.; Tworek, M.; Kurek, M. Atrazine, triketone herbicides, and their degradation products in sediment, soil and surface water samples in Poland. *Environ. Sci. Pollut. Res. Int.,* **2017,** *24*(1), 644-658.
[http://dx.doi.org/10.1007/s11356-016-7798-3] [PMID: 27743329]

[178] Zhou, T.; Ding, J.; Wang, Q.; Xu, Y.; Wang, B.; Zhao, L.; Ding, H.; Chen, Y.; Ding, L. Microwave-assisted rapid preparation of monodisperse superhydrophilic resin microspheres as adsorbent for triazines in fruit juices. *Talanta,* **2018,** *179,* 734-741.
[http://dx.doi.org/10.1016/j.talanta.2017.12.003] [PMID: 29310301]

[179] Wu, L.; Hu, M.; Li, Z.; Song, Y.; Yu, C.; Zhang, Y.; Zhang, H.; Yu, A.; Ma, Q.; Wang, Z. Determination of triazine herbicides in fresh vegetables by dynamic microwave-assisted extraction coupled with homogeneous ionic liquid microextraction high performance liquid chromatography. *Anal. Bioanal. Chem.,* **2015,** *407*(6), 1753-1762.
[http://dx.doi.org/10.1007/s00216-014-8393-4] [PMID: 25542578]

[180] Li, N.; Wu, L.; Nian, L.; Song, Y.; Lei, L.; Yang, X.; Wang, K.; Wang, Z.; Zhang, L.; Zhang, H.; Yu, A.; Zhang, Z. Dynamic microwave assisted extraction coupled with dispersive micro-solid-phase extraction of herbicides in soybeans. *Talanta,* **2015,** *142,* 43-50.
[http://dx.doi.org/10.1016/j.talanta.2015.04.038] [PMID: 26003690]

[181] Wang, H.; Li, G.; Zhang, Y.; Chen, H.; Zhao, Q.; Song, W.; Xu, Y.; Jin, H.; Ding, L. Determination of triazine herbicides in cereals using dynamic microwave-assisted extraction with solidification of floating organic drop followed by high-performance liquid chromatography. *J. Chromatogr. A,* **2012,** *1233,* 36-43.
[http://dx.doi.org/10.1016/j.chroma.2012.02.034] [PMID: 22402128]

[182] Cheng, J.; Liu, M.; Zhang, X.; Ding, L.; Yu, Y.; Wang, X.; Jin, H.; Zhang, H. Determination of triazine herbicides in sheep liver by microwave-assisted extraction and high performance liquid chromatography. *Anal. Chim. Acta,* **2007,** *590*(1), 34-39.
[http://dx.doi.org/10.1016/j.aca.2007.03.017] [PMID: 17416220]

[183] Vryzas, Z.; Papadopoulou-Mourkidou, E. Determination of triazine and chloroacetanilide herbicides in soils by microwave-assisted extraction (MAE) coupled to gas chromatographic analysis with either GC-NPD or GC-MS. *J. Agric. Food Chem.,* **2002,** *50*(18), 5026-5033.
[http://dx.doi.org/10.1021/jf020176f] [PMID: 12188602]

[184] Usui, K.; Hayashizaki, Y.; Hashiyada, M.; Funayama, M. Rapid drug extraction from human whole blood using a modified QuEChERS extraction method. *Leg. Med. (Tokyo),* **2012,** *14*(6), 286-296.
[http://dx.doi.org/10.1016/j.legalmed.2012.04.008] [PMID: 22682428]

[185] Vudathala, D.; Cummings, M.; Murphy, L. Analysis of multiple anticoagulant rodenticides in animal blood and liver tissue using principles of QuEChERS method. *J. Anal. Toxicol.,* **2010,** *34*(5), 273-279.
[http://dx.doi.org/10.1093/jat/34.5.273] [PMID: 20529461]

[186] Lesueur, C.; Gartner, M.; Mentler, A.; Fuerhacker, M. Comparison of four extraction methods for the analysis of 24 pesticides in soil samples with gas chromatography-mass spectrometry and liquid chromatography-ion trap-mass spectrometry. *Talanta,* **2008,** *75*(1), 284-293.
[http://dx.doi.org/10.1016/j.talanta.2007.11.031] [PMID: 18371880]

[187] Mei, M.; Du, Z-X.; Cen, Y. QuEChERS-Ultra-Performance Liquid Chromatography Tandem Mass Spectrometry for Determination of Five Currently Used Herbicides. *Chin. J. Anal. Chem.,* **2011,** *39*(11), 1659-1664.

[http://dx.doi.org/10.1016/S1872-2040(10)60482-3]

[188] Nagel, T.G. The QuEChERS method - a new approach in pesticide analysis of soils. *J. Hortic. Sci. Biotechnol.,* **2009**, *13*, 391-393.

[189] Muhammad, M.; Jan, M.R.; Shah, J.; Ara, B.; Akhtar, S.; Rahman, H.U. Evaluation and statistical analysis of the modified QuEChERS method for the extraction of pinoxaden from environmental and agricultural samples. *J. Anal. Sci. Technol.,* **2017**, *8*(1), 12.
[http://dx.doi.org/10.1186/s40543-017-0123-z]

[190] Li, J.; Gu, Y.; Xue, J.; Jin, H.Y.; Ma, S. Analysis and Risk Assessment of Pesticide Residues in a Chinese Herbal Medicine, Lonicera japonica Thunb. *Chromatographia,* **2017**, *80*(3), 503-512.
[http://dx.doi.org/10.1007/s10337-017-3243-0]

[191] Dong, B.; Qian, W.; Hu, J. Dissipation kinetics and residues of florasulam and tribenuron-methyl in wheat ecosystem. *Chemosphere,* **2015**, *120*, 486-491.
[http://dx.doi.org/10.1016/j.chemosphere.2014.09.016] [PMID: 25268470]

[192] Cui, K.; Wu, X.; Zhu, L.; Zhang, Y.; Dai, G.; Cao, J.; Xu, J.; Dong, F.; Liu, X.; Zheng, Y. Development and establishment of a QuEChERS-based extraction method for determining tembotrione and its metabolite AE 1417268 in corn, corn oil and certain animal-origin foods by HPLC-MS/MS. *Food Addit. Contam. Part A Chem. Anal. Control Expo. Risk Assess.,* **2020**, *37*(10), 1678-1686.
[http://dx.doi.org/10.1080/19440049.2020.1787526] [PMID: 32787691]

[193] Pang, K.; Hu, J. Simultaneous analysis and dietary exposure risk assessment of fomesafen, clomazone, clethodim and its two metabolites in soybean ecosystem. *Int. J. Environ. Res. Public Health,* **2020**, *17*(6), 1951.
[http://dx.doi.org/10.3390/ijerph17061951] [PMID: 32191999]

[194] Demoliner, A.; Caldas, S.S.; Costa, F.P.; Gonçalves, F.F.; Clementin, R.M.; Milani, M.R.; Primel, E.G. Development and validation of a method using SPE and LC-ESI-MS-MS for the determination of multiple classes of pesticides and metabolites in water samples. *J. Braz. Chem. Soc.,* **2010**, *21*(8), 1424-1433.
[http://dx.doi.org/10.1590/S0103-50532010000800003]

[195] Srivastava, M; Suyal, A. Persistence behavior of penoxsulam herbicide in two different soils. *Bull. Environ. Contam. Toxicol.,* **2017**, *99*, 470-474.
[http://dx.doi.org/10.1007/s00128-017-2171-x] [PMID: 28875291]

[196] Lee, Y.J.; Rahman, M.M.; Abd El-Aty, A.M.; Choi, J.H.; Chung, H.S.; Kim, S.W.; Abdel-Aty, A.M.; Shin, H.C.; Shim, J.H. Detection of three herbicide, and one metabolite, residues in brown rice and rice straw using various versions of the QuEChERS method and liquid chromatography-tandem mass spectrometry. *Food Chem.,* **2016**, *210*, 442-450.
[http://dx.doi.org/10.1016/j.foodchem.2016.05.005] [PMID: 27211669]

[197] Rani, N.; Duhan, A.; Tomar, D. Ultimate fate of herbicide tembotrione and its metabolite TCMBA in soil. *Ecotoxicol. Environ. Saf.,* **2020**, *203*, 111023.
[http://dx.doi.org/10.1016/j.ecoenv.2020.111023] [PMID: 32888592]

[198] Ramos, J.J.; Rial-Otero, R.; Ramos, L.; Capelo, J.L. Ultrasonic-assisted matrix solid-phase dispersion as an improved methodology for the determination of pesticides in fruits. *J. Chromatogr. A,* **2008**, *1212*(1-2), 145-149.
[http://dx.doi.org/10.1016/j.chroma.2008.10.028] [PMID: 18952216]

[199] Souza Caldas, S.; Marian Bolzan, C.; Jaime de Menezes, E.; Laura Venquiaruti Escarrone, A.; de Martinez Gaspar Martins, C.; Bianchini, A.; Gilberto Primel, E. A vortex-assisted MSPD method for the extraction of pesticide residues from fish liver and crab hepatopancreas with determination by GC-MS. *Talanta,* **2013**, *112*, 63-68.
[http://dx.doi.org/10.1016/j.talanta.2013.03.054] [PMID: 23708538]

[200] Fotouhi, M.; Seidi, S.; Shanehsaz, M.; Naseri, M.T. Magnetically assisted matrix solid phase

dispersion for extraction of parabens from breast milks. *J. Chromatogr. A,* **2017**, *1504*, 17-26.
[http://dx.doi.org/10.1016/j.chroma.2017.05.009] [PMID: 28502468]

[201] Albero, B.; Tadeo, J.L.; Pérez, R.A. Ultrasound-assisted extraction of organic contaminants. *Trends Analyt. Chem.*, **2019**, *118*, 739-750.
[http://dx.doi.org/10.1016/j.trac.2019.07.007]

[202] Shi, X.; Sun, A.; Wang, Q.; Hengel, M.; Shibamoto, T. Rapid multi-residue analysis of herbicides with endocrine-disrupting properties in environmental water samples using ultrasound-assisted dispersive liquid-liquid microextraction and gas chromatography-mass spectrometry. *Chromatographia,* **2018**, *81*(7), 1071-1083.
[http://dx.doi.org/10.1007/s10337-018-3530-4]

[203] Ramos, J.J.; Rial-Otero, R.; Ramos, L.; Capelo, J.L. Ultrasonic-assisted matrix solid-phase dispersion as an improved methodology for the determination of pesticides in fruits. *J. Chromatogr. A,* **2008**, *1212*(1-2), 145-149.
[http://dx.doi.org/10.1016/j.chroma.2008.10.028] [PMID: 18952216]

[204] Souza Caldas, S.; Marian Bolzan, C.; Jaime de Menezes, E.; Laura Venquiaruti Escarrone, A.; de Martinez Gaspar Martins, C.; Bianchini, A.; Gilberto Primel, E. A vortex-assisted MSPD method for the extraction of pesticide residues from fish liver and crab hepatopancreas with determination by GC-MS. *Talanta,* **2013**, *112*, 63-68.
[http://dx.doi.org/10.1016/j.talanta.2013.03.054] [PMID: 23708538]

[205] Soares, K.L.; Cerqueira, M.B.R.; Caldas, S.S.; Primel, E.G. Evaluation of alternative environmentally friendly matrix solid phase dispersion solid supports for the simultaneous extraction of 15 pesticides of different chemical classes from drinking water treatment sludge. *Chemosphere,* **2017**, *182*, 547-554.
[http://dx.doi.org/10.1016/j.chemosphere.2017.05.062] [PMID: 28525867]

[206] Rombaldi, C.; de Oliveira Arias, J.L.; Hertzog, G.I.; Caldas, S.S.; Vieira, J.P.; Primel, E.G. New environmentally friendly MSPD solid support based on golden mussel shell: characterization and application for extraction of organic contaminants from mussel tissue. *Anal. Bioanal. Chem.,* **2015**, *407*(16), 4805-4814.
[http://dx.doi.org/10.1007/s00216-015-8686-2] [PMID: 25925855]

[207] Cai, J.; Gao, Y.; Zhu, X.; Su, Q. Matrix solid phase dispersion-Soxhlet simultaneous extraction clean-up for determination of organochlorine pesticide residues in tobacco. *Anal. Bioanal. Chem.,* **2005**, *383*(5), 869-874.
[http://dx.doi.org/10.1007/s00216-005-0076-8] [PMID: 16211380]

[208] Majlesi, M.; Massoudinejad, M.; Hosainzadeh, F.; Fattahi, N. Simultaneous separation and preconcentration of phosalone and chlorpyrifos in fresh vegetables using ultrasound-assisted dispersive liquid-liquid microextraction and high performance liquid chromatography. *Anal. Methods,* **2016**, *8*(18), 3795-3801.
[http://dx.doi.org/10.1039/C6AY00144K]

[209] Shi, X.; Sun, A.; Wang, Q.; Hengel, M.; Shibamoto, T. Rapid multi-residue analysis of herbicides with endocrine-disrupting properties in environmental water samples using ultrasound-assisted dispersive liquid-liquid microextraction and gas chromatography-mass spectrometry. *Chromatographia,* **2018**, *81*(7), 1071-1083.
[http://dx.doi.org/10.1007/s10337-018-3530-4]

[210] Catrinck, T.C.P.G.; Dias, A.; Aguiar, M.C.S.; Silvério, F.O.; Fidêncio, P.H.; Pinho, G.P. A Simple and Efficient Method for Derivatization of Glyphosate and AMPA Using 9-Fluorenylmethyl Chloroformate and Spectrophotometric Analysis. *J. Braz. Chem. Soc.,* **2014**, *25*, 1194-1199.
[http://dx.doi.org/10.5935/0103-5053.20140096]

[211] Shah, J.; Jan, M.R.; Muhammad, M.; Shehzad, F.N. Flow injection spectrophotometric determination of fenoxaprop-p-ethyl herbicide in different grain samples after derivatization. *J. Braz. Chem. Soc.,* **2010**, *21*(10), 1923-1928.
[http://dx.doi.org/10.1590/S0103-50532010001000018]

[212]   Ara, B.; Shah, J.; Rasul Jan, M.; Muhammad, M. Spectrophotometric determination of metribuzin herbicide with p -dimethylamino-benzaldehyde using factorial designs for optimization of experimental variables. *J. Saudi Chem. Soc.,* **2016**, *20*, S566-S572.
[http://dx.doi.org/10.1016/j.jscs.2013.03.006]

[213]   Khan, K.A.; Gulab, H.; Haider, F. Development of spectrophotometric method for the determination of azoxystrobin fungicide after derivatization. *Biochem. Anal. Biochem.,* **2018**, *7*, 2.

[214]   Khan, K.A.; Haider, F. Spectrophotometric determination and commercial formulation of tebuconazole fungicide after derivatization. *Chem. Sci. J.,* **2016**, *7*, 2.

[215]   Kaur, I.; Gaba, S.; Kaur, S.; Kumar, R.; Chawla, J. Spectrophotometric determination of triclosan based on diazotization reaction: response surface optimization using Box-Behnken design. *Water Sci. Technol.,* **2018**, *77*(9), 2204-2212.
[http://dx.doi.org/10.2166/wst.2018.123] [PMID: 29757172]

[216]   Alrahman, K.F.; Elbashir, A.A.; Ahmed, H.E. Development and validation of spectrophotometric method for determination of oxyfluorfen herbicide residues. *Med. Chem.,* **2015**, *5*, 383-387.

[217]   Bernal, J.; Martin, M.T.; Soto, M.E.; Nozal, M.J.; Marotti, I.; Dinelli, G.; Bernal, J.L. Development and application of a liquid chromatography-mass spectrometry method to evaluate the glyphosate and aminomethylphosphonic acid dissipation in maize plants after foliar treatment. *J. Agric. Food Chem.,* **2012**, *60*(16), 4017-4025.
[http://dx.doi.org/10.1021/jf3006504] [PMID: 22480367]

[218]   Druart, C.; Delhomme, O.; de Vaufleury, A.; Ntcho, E.; Millet, M. Optimization of extraction procedure and chromatographic separation of glyphosate, glufosinate and aminomethylphosphonic acid in soil. *Anal. Bioanal. Chem.,* **2011**, *399*(4), 1725-1732.
[http://dx.doi.org/10.1007/s00216-010-4468-z] [PMID: 21153586]

[219]   Mendy, A.; Thiaré, D.D.; Sambou, S.; Khonté, A.; Coly, A.; Gaye-Seye, M.D.; Delattre, F.; Tine, A. New method for the determination of metolachlor and buprofezin in natural water using orthophthalaldehyde by thermochemically-induced fluorescence derivatization (TIFD). *Talanta,* **2016**, *151*, 202-208.
[http://dx.doi.org/10.1016/j.talanta.2016.01.036] [PMID: 26946028]

[220]   Coly, A.; Aaron, J.J. Sensitive and rapid flow injection analysis of sulfonylurea herbicides in water with micellar-enhanced photochemically induced fluorescence detection. *Anal. Chim. Acta,* **1999**, *392*(2-3), 255-264.
[http://dx.doi.org/10.1016/S0003-2670(99)00229-9]

[221]   Pal, A. Bandyopadhyay. fluorometric determination of trichloroacetic acid and its application in water sample analysis. *Indian J. Chem. Technol.,* **1997**, *4*, 253-255.

[222]   Muhammad, M.; Shah, J.; Jan, M.R.; Ara, B.; Mahabat Khan, M.; Jan, A. Spectrofluorimetric method for quantification of triazine herbicides in agricultural matrices. *Anal. Sci.,* **2016**, *32*(3), 313-316.
[http://dx.doi.org/10.2116/analsci.32.313] [PMID: 26960611]

[223]   Attimarad, M.; Mueen Ahmed, K.K.; Aldhubaib, B.E.; Harsha, S. High-performance thin layer chromatography: A powerful analytical technique in pharmaceutical drug discovery. *Pharm. Methods,* **2011**, *2*(2), 71-75.
[http://dx.doi.org/10.4103/2229-4708.84436] [PMID: 23781433]

[224]   Haskis, P.; Mantzos, N.; Hela, D.; Patakioutas, G.; Konstantinou, I. Effect of biochar on the mobility and photodegradation of metribuzin and metabolites in soil–biochar thin-layer chromatography plates. *Int. J. Environ. Anal. Chem.,* **2019**, *99*(4), 310-327.
[http://dx.doi.org/10.1080/03067319.2019.1597863]

[225]   Daneshfar, A.; Khezeli, T. Headspace solid phase microextraction of nicotine using thin layer chromatography plates modified with carbon dots. *Mikrochim. Acta,* **2015**, *182*(1-2), 209-216.
[http://dx.doi.org/10.1007/s00604-014-1318-2]

[226] Deng, H.; Feng, D.; He, J.; Li, F.; Yu, H.; Ge, C. Influence of biochar amendments to soil on the mobility of atrazine using sorption-desorption and soil thin-layer chromatography. *Ecol. Eng.,* **2017**, *99*, 381-390.
[http://dx.doi.org/10.1016/j.ecoleng.2016.11.021]

[227] Kavrakovski, Z.S.; Rafajlovska, V.G. Development and validation of thin layer chromatography method for simultaneous determination of seven chlorophenoxy and benzoic acid herbicides in water. *J. Anal. Chem.,* **2015**, *70*(8), 995-1000.
[http://dx.doi.org/10.1134/S1061934815080122]

[228] Cobzac, S.; Olah, N.K.; Gocan, S. TLC determination of triazinic pesticides from soils — A comparative study of some extraction methods. *J. Planar Chromatogr. Mod. TLC,* **2012**, *25*(2), 97-100.
[http://dx.doi.org/10.1556/JPC.25.2012.2.1]

[229] Broszat, M.; Ernst, H.; Spangenberg, B. A Simple Method for Quantification Triazine Herbicides Using Thin-Layer Chromatography and A CCD Camera. *J. Liq. Chromatogr. Relat. Technol.,* **2010**, *33*(7-8), 948-956.
[http://dx.doi.org/10.1080/10826071003766245]

[230] Pawar, U.D.; Pawar, C.D.; Kulkarni, U.K.; Pardeshi, R.K. Development method of high-performance thin-layer chromatographic detection of synthetic organophosphate insecticide profenofos in visceral samples. *J. Planar Chromatogr. Mod. TLC,* **2020**, *33*(2), 203-206.
[http://dx.doi.org/10.1007/s00764-020-00015-2]

[231] Ozdemir, S.; Bekler, F.M.; Okumus, V.; Dundar, A.; Kilinc, E. Biosorption of 2,4-d, 2,4-DP, and 2,4-DB from aqueous solution by using thermophilic anoxybacillus flavithermus and analysis by high-performance thin layer chromatography: Equilibrium and kinetic studies. *Environ. Prog. Sustain. Energy,* **2012**, *31*(4), 544-552.
[http://dx.doi.org/10.1002/ep.10576]

[232] Espinoza, J.; Báez, M. Determination of atrazine in aqueous soil extracts by high performance thin-layer chromatography. *J. Chil. Chem. Soc.,* **2003**, *48*(1), 1.
[http://dx.doi.org/10.4067/S0717-97072003000100004]

[233] Ovchinnikova, O.S.; Van Berkel, G.J. Thin-layer chromatography and mass spectrometry coupled using proximal probe thermal desorption with electrospray or atmospheric pressure chemical ionization. *Rapid Commun. Mass Spectrom.,* **2010**, *24*(12), 1721-1729.
[http://dx.doi.org/10.1002/rcm.4551] [PMID: 20499315]

[234] Si, Y.; Wang, S.; Zhou, D.; Chen, H. Adsorption and photo-reactivity of bensulfuron-methyl on homoionic clays. *Clays Clay Miner.,* **2004**, *52*(6), 742-748.
[http://dx.doi.org/10.1346/CCMN.2004.0520609]

[235] Kaur, I.; Gaba, S.; Kaur, S.; Kumar, R.; Chawla, J. Spectrophotometric determination of triclosan based on diazotization reaction: response surface optimization using Box-Behnken design. *Water Sci. Technol.,* **2018**, *77*(9), 2204-2212.
[http://dx.doi.org/10.2166/wst.2018.123] [PMID: 29757172]

[236] Kaur, P; Gupta, R C; Dey, A; Pandey, D K. Simultaneous quantification of oleanolic acid, ursolic acid, betulinic acid and lupeol in different populations of five Swertia species by using HPTLCdensitometry: Comparison of different extraction methods and solvent selection. *Industrial Crops Products,* **2019**, *130*, 537-546.

[237] Sun, Y.; Liu, P.F.; Wang, D.; Li, J.Q.; Cao, Y.S. Determination of amitrole in environmental water samples with precolumn derivatization by high-performance liquid chromatography. *J. Agric. Food Chem.,* **2009**, *57*(11), 4540-4544.
[http://dx.doi.org/10.1021/jf900601f] [PMID: 19425574]

[238] N L. Determination of glyphosate residues in water by liquid chromatography. *China Measure Technol,* **2007**, *33*, 114-116.

[239]   L L. Determination of organophosphorus herbicide in water by high Performance liquid chromatography pre-column derivatization. *Environ Monit China,* **2009**, *25*, 35-38.

[240]   WJ G. Determination of glyphosate in water by high performance liquid chromatography-fluorescence detection with pre-column derivatization and solid-phase extraction. *Chin J Health Lab Technol,* **2014**, *24*, 2599-2601.

[241]   Fang, F.; Hui, X.; Qing, W.R.; Ning, L.X.; Rong, X.; Chun, L.S.; Kun, L.B. Determination of glyphosate by high performance liquid chromatography with o-nitrobenzenesulfonyl chloride as derivatization reagent. *Chin J Instrum Anal,* **2011**, *30*, 683-686.

[242]   Youbin, S.; Ying, S.Z.; Xia, C.F. Determination of glyphosate in soil by high performance liquid chromatography after derivatization with p-toluenesulphonyl chloride. *Chin J Anhui Agric Univ,* **2009**, *36*, 136-139.

[243]   Liu, L.; Xia, L.; Guo, C.; Wu, C.; Chen, G.; Li, G.; Sun, Z.; You, J. A sensitive and efficient method for the determination of 8 chlorophenoxy acid herbicides in crops by dispersive liquid-liquid microextraction and HPLC with fluorescence detection and identification by MS. *Anal. Methods,* **2016**, *8*(17), 3536-3544.
[http://dx.doi.org/10.1039/C5AY03412D]

[244]   Hu, Y.S.; Zhao, Y.Q.; Sorohan, B. Removal of glyphosate from aqueous environment by adsorption using water industrial residual. *Desalination,* **2011**, *271*(1-3), 150-156.
[http://dx.doi.org/10.1016/j.desal.2010.12.014]

[245]   Pei, M.Q.; Lai, J. Qualitative and quantitative analysis of glyphosate. *Chin J Guangdong Police Sci Technol,* **2004**, *1*, 14-15.

[246]   Tsunoda, N. Simultaneous determination of the herbicides glyphosate, glufosinate and bialaphos and their metabolites by capillary gas chromatography—ion-trap mass spectrometry. *J. Chromatogr. A,* **1993**, *637*(2), 167-173.
[http://dx.doi.org/10.1016/0021-9673(93)83209-B]

[247]   Kudzin, Z.H.; Gralak, D.K.; Drabowicz, J.; Łuczak, J. Novel approach for the simultaneous analysis of glyphosate and its metabolites. *J. Chromatogr. A,* **2002**, *947*(1), 129-141.
[http://dx.doi.org/10.1016/S0021-9673(01)01603-X] [PMID: 11873992]

[248]   Lou, Z.Y.; Zhu, G.N.; Wu, H.M. Study on the detection method of glyphosate in pond water. *Chin J Ningbo Acad,* **2001**, *13*, 142-145.

[249]   Shamsipur, M.; Fattahi, N.; Pirsaheb, M.; Sharafi, K. Simultaneous preconcentration and determination of 2,4-D, alachlor and atrazine in aqueous samples using dispersive liquid-liquid microextraction followed by high-performance liquid chromatography ultraviolet detection. *J. Sep. Sci.,* **2012**, *35*(20), 2718-2724.
[http://dx.doi.org/10.1002/jssc.201200424] [PMID: 22997055]

[250]   Beldean-Galea, M.S.; Dragus, A.; Mihaiescu, R.; Mihaiescu, T.; Ristoiu, D. Assessing impacts of triazine pesticides use in agriculture over the well water quality. *Environ. Eng. Manag. J.,* **2012**, *11*(2), 319-323.
[http://dx.doi.org/10.30638/eemj.2012.041]

[251]   Guo, C.; Hu, J.Y.; Chen, X.Y.; Li, J.Z. Analysis of imazaquin in soybeans by solid-phase extraction and high-performance liquid chromatography. *Bull. Environ. Contam. Toxicol.,* **2008**, *80*(2), 173-177.
[http://dx.doi.org/10.1007/s00128-007-9340-2] [PMID: 18183337]

[252]   Wang, C.; Ji, S.; Wu, Q.; Wu, C.; Wang, Z. Determination of triazine herbicides in environmental samples by dispersive liquid-liquid microextraction coupled with high performance liquid chromatography. *J. Chromatogr. Sci.,* **2011**, *49*(9), 689-694.
[http://dx.doi.org/10.1093/chrsci/49.9.689] [PMID: 22586245]

[253]   Rodríguez-González, N.; González-Castro, M-J.; Beceiro-González, E.; Muniategui-Lorenzo, S. Development of a matrix solid phase dispersion methodology for the determination of triazine

herbicides in marine sediments. *Microchem. J.,* **2017**, *133*, 137-143.
[http://dx.doi.org/10.1016/j.microc.2017.03.022]

[254] Barchanska, H.; Babilas, B.; Gluzicka, K.; Zralek, D.; Baranowska, I. Rapid determination of mesotrione, atrazine and its main degradation products in selected plants by MSPD - HPLC and indirect estimation of herbicides phytotoxicity by chlorophyll quantification. Intern J Environ. *Anal. Chem.,* **2013**, *94*(2), 99-114.

[255] Liang, P.; Wang, J.; Liu, G.; Guan, J. Determination of sulfonylurea herbicides in food crops by matrix solid-phase dispersion extraction coupled with high-performance liquid chromatography. *Food Anal. Methods,* **2014**, *7*(7), 1530-1535.
[http://dx.doi.org/10.1007/s12161-013-9784-4]

[256] Lee, Y.J.; Rahman, M.M.; Abd El-Aty, A.M.; Choi, J.H.; Chung, H.S.; Kim, S.W.; Abdel-Aty, A.M.; Shin, H.C.; Shim, J.H. Detection of three herbicide, and one metabolite, residues in brown rice and rice straw using various versions of the quechers method and liquid chromatography-tandem mass spectrometry. *Food Chem.,* **2016**, *210*, 442-450.
[http://dx.doi.org/10.1016/j.foodchem.2016.05.005] [PMID: 27211669]

[257] Pang, N.; Wang, T.; Hu, J. Method validation and dissipation kinetics of four herbicides in maize and soil using QuEChERS sample preparation and liquid chromatography tandem mass spectrometry. *Food Chem.,* **2016**, *190*, 793-800.
[http://dx.doi.org/10.1016/j.foodchem.2015.05.081] [PMID: 26213040]

[258] Li, Z.; Bai, S.; Hou, M.; Wang, C.; Wang, Z. Magnetic Graphene Nanoparticles for the Preconcentration of Chloroacetanilide Herbicides from Water Samples Prior to Determination by GC-ECD. *Anal. Lett.,* **2013**, *46*(6), 1012-1024.
[http://dx.doi.org/10.1080/00032719.2012.745086]

[259] Hu, M.; Chen, H.; Jiang, Y.; Zhu, H. Headspace single-drop microextraction coupled with gas chromatography electron capture detection of butanone derivative for determination of iodine in milk powder and urine. *Chem. Pap.,* **2013**, *67*(10), 1255-1261.
[http://dx.doi.org/10.2478/s11696-013-0391-z]

[260] Wielgomas, B.; Czarnowski, W. Headspace single-drop microextraction and GC-ECD determination of chlorpyrifos-ethyl in rat liver. *Anal. Bioanal. Chem.,* **2008**, *390*(7), 1933-1941.
[http://dx.doi.org/10.1007/s00216-008-1831-4] [PMID: 18299820]

[261] Furlani, R.P.Z.; Marcilio, K.M.; Leme, F.M.; Tfouni, S.A.V. Analysis of pesticide residues in sugarcane juice using QuEChERS sample preparation and gas chromatography with electron capture detection. *Food Chem.,* **2011**, *126*(3), 1283-1287.
[http://dx.doi.org/10.1016/j.foodchem.2010.11.074]

[262] Berrada, H.; Font, G.; Moltó, J.C. Application of solid-phase microextraction for determining phenylurea herbicides and their homologous anilines from vegetables. *J. Chromatogr. A,* **2004**, *1042*(1-2), 9-14.
[http://dx.doi.org/10.1016/j.chroma.2004.05.017] [PMID: 15296383]

[263] Avramides, E.J.; Gkatsos, S. A multiresidue method for the determination of insecticides and triazine herbicides in fresh and processed olives. *J. Agric. Food Chem.,* **2007**, *55*(3), 561-565.
[http://dx.doi.org/10.1021/jf062826b] [PMID: 17263441]

[264] Ángeles García, M.; Santaeufemia, M.; Julia Melgar, M. Triazine residues in raw milk and infant formulas from Spanish northwest, by a diphasic dialysis extraction. *Food Chem. Toxicol.,* **2012**, *50*(3-4), 503-510.
[http://dx.doi.org/10.1016/j.fct.2011.11.019] [PMID: 22137903]

[265] Araujo, L.; Prieto, A.; Troconis, M.; Urribarri, G.; Sandrea, W.; Mercado, J. Determination of acidic herbicides in water samples by in situ derivatization, single drop microextraction and gas chromatography-mass spectrometry. *J. Braz. Chem. Soc.,* **2011**, *22*(12), 2350-2354.
[http://dx.doi.org/10.1590/S0103-50532011001200016]

[266] Shen, G.; Lee, H.K. Hollow fiber-protected liquid-phase microextraction of triazine herbicides. *Anal. Chem.*, **2002**, *74*(3), 648-654.
[http://dx.doi.org/10.1021/ac010561o] [PMID: 11838687]

[267] Ramos, J.J.; Rial-Otero, R.; Ramos, L.; Capelo, J.L. Ultrasonic-assisted matrix solid-phase dispersion as an improved methodology for the determination of pesticides in fruits. *J. Chromatogr. A,* **2008**, *1212*(1-2), 145-149.
[http://dx.doi.org/10.1016/j.chroma.2008.10.028] [PMID: 18952216]

[268] Li, Z.; Li, Y.; Liu, X.; Li, X.; Zhou, L.; Pan, C. Multiresidue analysis of 58 pesticides in bean products by disposable pipet extraction (DPX) cleanup and gas chromatography-mass spectrometry determination. *J. Agric. Food Chem.*, **2012**, *60*(19), 4788-4798.
[http://dx.doi.org/10.1021/jf300234d] [PMID: 22394480]

[269] Bielawski, D.; Ostrea, E., Jr; Posecion, N., Jr; Corrion, M.; Seagraves, J. Detection of several classes of pesticides and metabolites in meconium by gas chromatography-mass spectrometry. *Chromatographia*, **2005**, *62*(11-12), 623-629.
[http://dx.doi.org/10.1365/s10337-005-0668-7] [PMID: 17664958]

[270] Carrai, P.; Nucci, L.; Pergola, F. Polarographic behaviour of alachlor application to analytical determination. *Anal. Lett.*, **1992**, *25*(1), 163-172.
[http://dx.doi.org/10.1080/00032719208020017]

[271] Lippolis, M.T.; Concialini, V. Differential pulse polarographic determination of the herbicides atrazine, prometrine and simazine. *Talanta*, **1988**, *35*(3), 235-236.
[http://dx.doi.org/10.1016/0039-9140(88)80072-9] [PMID: 18964502]

[272] Walcarius, A.; Lamberts, L. Square wave voltammetric determination of paraquat and diquat in aqueous solution. *J. Electroanal. Chem. (Lausanne)*, **1996**, *406*(1-2), 59-68.
[http://dx.doi.org/10.1016/0022-0728(95)04385-3]

[273] Walcarius, A.; Lamberts, L. Square wave voltammetric determination of paraquat and diquat in aqueous solution. *J. Electroanal. Chem. (Lausanne)*, **1996**, *406*(1-2), 59-68.
[http://dx.doi.org/10.1016/0022-0728(95)04385-3]

[274] Kotouček, M.; Opravilová, M. Voltammetric behaviour of some nitropesticides at the mercury drop electrode. *Anal. Chim. Acta*, **1996**, *329*(1-2), 73-81.
[http://dx.doi.org/10.1016/0003-2670(96)00133-X]

[275] Garrido, E.M.; Lima, J.L.F.C.; Delerue-Matos, C.; Borges, F.; Silva, A.M.S.; Piedade, J.A.P.; Oliveira Brett, A.M. Electrochemical and spectroscopic studies of the oxidation mechanism of the herbicide propanil. *J. Agric. Food Chem.*, **2003**, *51*(4), 876-879.
[http://dx.doi.org/10.1021/jf025957v] [PMID: 12568542]

[276] Ignjatović, L.M.; Marković, D.A.; Veselinović, D.S.; Bešić, B.R. Polarographic behavior and determination of some S-triazine herbicides. *Electroanalysis,* **1993**, *5*(5-6), 529-533.
[http://dx.doi.org/10.1002/elan.1140050525]

[277] Concialini, V.; Lippolis, M.T.; Galletti, G.C. Preliminary studies for the differential-pulse polarographic determination of a new class of herbicides: sulphonylureas. *Analyst (Lond.),* **1989**, *114*(12), 1617.
[http://dx.doi.org/10.1039/an9891401617]

[278] El Harmoudi, H.; Achak, M.; Farahi, A.; Lahrich, S.; El Gaini, L.; Abdennouri, M.; Bouzidi, A.; Bakasse, M.; El Mhammedi, M.A. Sensitive determination of paraquat by square wave anodic stripping voltammetry with chitin modified carbon paste electrode. *Talanta*, **2013**, *115*, 172-177.
[http://dx.doi.org/10.1016/j.talanta.2013.04.002] [PMID: 24054575]

[279] Lima, A.C.A.; Silva, E.G.; Goulart, M.O.F.; Tonholo, J.; Silva, T.T.; Abreu, F.C. Electrochemical behavior of metribuzin on a glassy carbon electrode in an aqueous medium including quantitative studies by anodic stripping voltammetry. *J. Braz. Chem. Soc.*, **2009**, *20*(9), 1698-1704.

[http://dx.doi.org/10.1590/S0103-50532009000900019]

[280] Dzantiev, B.; Yazynina, E.V.; Zherdev, A.V.; Plekhanova, Y.V.; Reshetilov, A.N.; Chang, S.C.; McNeil, C.J. Determination of the herbicide chlorsulfuron by amperometric sensor based on separation-free bienzyme immunoassay. *Sens. Actuators B Chem.,* **2004**, *98*(2-3), 254-261.
[http://dx.doi.org/10.1016/j.snb.2003.10.021]

[281] Mani, V.; Devasenathipathy, R.; Chen, S.M.; Wu, T.Y.; Kohilarani, K. High-performance electrochemical amperometric sensors for the sensitive determination of phenyl urea herbicides diuron and fenuron. *Ionics,* **2015**, *21*(9), 2675-2683.
[http://dx.doi.org/10.1007/s11581-015-1459-2]

[282] Chang, P.L.; Hsieh, M.M.; Chiu, T.C. Recent advances in the determination of pesticides in environmental samples by capillary electrophoresis. *Int. J. Environ. Res. Public Health,* **2016**, *13*(4), 409.
[http://dx.doi.org/10.3390/ijerph13040409] [PMID: 27070634]

[283] Dong, Y.; Guo, D.; Cui, H.; Li, X.; He, Y. Magnetic solid phase extraction of glyphosate and aminomethylphosphonic acid in river water using Ti^{4+}-immobilized Fe_3O_4 nanoparticles by capillary electrophoresis. *Anal. Methods,* **2015**, *7*(14), 5862-5868.
[http://dx.doi.org/10.1039/C5AY00109A]

[284] Zhou, Q.; Mao, J.; Xiao, J.; Xie, G. Uses of ionic liquid 1-butyl-3-methylimidazolium hexafluorophosphate as a good separation electrolyte for direct electrophoretic separation of quaternary ammonium herbicides. *J. Sep. Sci.,* **2010**, *33*(9), 1288-1293.
[http://dx.doi.org/10.1002/jssc.200900770] [PMID: 20187035]

[285] Tabani, H.; Fakhari, A.R.; Shahsavani, A.; Behbahani, M.; Salarian, M.; Bagheri, A.; Nojavan, S. Combination of graphene oxide-based solid phase extraction and electro membrane extraction for the preconcentration of chlorophenoxy acid herbicides in environmental samples. *J. Chromatogr. A,* **2013**, *1300*, 227-235.
[http://dx.doi.org/10.1016/j.chroma.2013.04.026] [PMID: 23688683]

[286] Zhou, L.; Su, P.; Deng, Y.; Yang, Y. Self-assembled magnetic nanoparticle supported zeolitic imidazolate framework-8: An efficient adsorbent for the enrichment of triazine herbicides from fruit, vegetables, and water. *J. Sep. Sci.,* **2017**, *40*(4), 909-918.
[http://dx.doi.org/10.1002/jssc.201601089] [PMID: 27987275]

[287] Kang, S.; Chang, N.; Zhao, Y.; Pan, C. Development of a method for the simultaneous determination of six sulfonylurea herbicides in wheat, rice, and corn by liquid chromatography-tandem mass spectrometry. *J. Agric. Food Chem.,* **2011**, *59*(18), 9776-9781.
[http://dx.doi.org/10.1021/jf2020073] [PMID: 21800900]

[288] Li, B.; Deng, X.; Guo, D.; Jin, S. [Determination of glyphosate and aminomethylphosphonic acid residues in foods using high performance liquid chromatography-mass spectrometry/mass spectrometry]. *Se Pu,* **2007**, *25*(4), 486-490.
[http://dx.doi.org/10.1016/S1872-2059(07)60017-0] [PMID: 17970103]

[289] Mei, M.; Du, Z.-X.; Cen, Y. QuEChERS-Ultra-Performance Liquid Chromatography Tandem Mass Spectrometry for Determination of Five Currently Used Herbicides. *Chin. J. Anal. Chem.,* **2011**, *39*(11), 1659-1664.
[http://dx.doi.org/10.1016/S1872-2040(10)60482-3]

[290] Qin, Z.; Jiang, Y.; Piao, H.; Li, J.; Tao, S.; Ma, P.; Wang, X.; Song, D.; Sun, Y. MIL-101(Cr)/MWCNTs-functionalized melamine sponges for solid-phase extraction of triazines from corn samples, and their subsequent determination by HPLC-MS/MS. *Talanta,* **2020**, *211*, 120676.
[http://dx.doi.org/10.1016/j.talanta.2019.120676] [PMID: 32070600]

[291] Wang, J; Leung, D; Chow, W; Wong, J W; Chang, J. UHPLC/ESI Q-Orbitrap Quantitation of 655 Pesticide Residues in Fruits and Vegetables – a Companion to an DATA Working Flow. *J ACOC Intern,* **2020**, *54*, 02-19.

[292] Xu, J.; Zhang, J.; Dong, F.; Liu, X.; Zhu, G.; Zheng, Y. A multiresidue analytical method for the detection of seven triazolopyrimidine sulfonamide herbicides in cereals, soybean and soil using the modified QuEChERS method and UHPLC-MS/MS. *Anal. Methods,* **2015**, *7*(23), 9791-9799.
[http://dx.doi.org/10.1039/C5AY01622C]

[293] Kalsi, N.K.; Kaur, P. Dissipation of bispyribac sodium in aridisols: Impact of soil type, moisture and temperature. *Ecotoxicol. Environ. Saf.,* **2019**, *170*, 375-382.
[http://dx.doi.org/10.1016/j.ecoenv.2018.12.005] [PMID: 30550967]

[294] Colazzo, M.; Pareja, L.; Cesio, M.V.; Heinzen, H. Multi-residue method for trace pesticide analysis in soils by LC-QQQ-MS/MS and its application to real samples. *Intern J Environ Anal Chem,* **2018**, 1-17.

[295] AMB D S and SC N D Q. Optimization and Validation of an Analytical Method for Determination of Herbicides Residues in Elephant Grass. *Austin Environ Sci,* **2020**, 5.

[296] Liang, L.; Wang, X.; Sun, Y.; Ma, P.; Li, X.; Piao, H.; Jiang, Y.; Song, D. Magnetic solid-phase extraction of triazine herbicides from rice using metal-organic framework MIL-101(Cr) functionalized magnetic particles. *Talanta,* **2018**, *179*, 512-519.
[http://dx.doi.org/10.1016/j.talanta.2017.11.017] [PMID: 29310269]

[297] AMB D S and SC N D Q. Optimization and Validation of an Analytical Method for Determination of Herbicides Residues in Elephant Grass. *Austin Environ Sci,* **2020**, 5.

[298] Djozan, D.; Ebrahimi, B.; Mahkam, M.; Farajzadeh, M.A. Evaluation of a new method for chemical coating of aluminum wire with molecularly imprinted polymer layer. Application for the fabrication of triazines selective solid-phase microextraction fiber. *Anal. Chim. Acta,* **2010**, *674*(1), 40-48.
[http://dx.doi.org/10.1016/j.aca.2010.06.006] [PMID: 20638497]

[299] Aramendia, M.; Borau, V.; Lafont, F.; Marinas, A.; Marinas, J.; Moreno, J.; Urbano, F. Determination of herbicide residues in olive oil by gas chromatography-tandem mass spectrometry. *Food Chem.,* **2007**, *105*(2), 855-861.
[http://dx.doi.org/10.1016/j.foodchem.2007.01.063]

[300] Lopez-Avila, V.; Roach, P.; Urdahl, R. Determination of Chlorophenoxy Acid Methyl Esters and Other Chlorinated Herbicides by GC High-resolution QTOFMS and Soft Ionization. *Anal. Chem. Insights,* **2015**, *10*, ACI.S21901.
[http://dx.doi.org/10.4137/ACI.S21901] [PMID: 25698878]

[301] Koblížek, M.; Malý, J.; Masojídek, J.; Komenda, J.; Kučera, T.; Giardi, M.T.; Mattoo, A.K.; Pilloton, R. A biosensor for the detection of triazine and phenylurea herbicides designed using Photosystem II coupled to a screen-printed electrode. *Biotechnol. Bioeng.,* **2002**, *78*(1), 110-116.
[http://dx.doi.org/10.1002/bit.10190] [PMID: 11857287]

[302] McArdle, F.A.; Persaud, K.C. Development of an enzyme-based biosensor for atrazine detection. *Analyst (Lond.),* **1993**, *118*(4), 419.
[http://dx.doi.org/10.1039/an9931800419]

[303] Besombes, J.L.; Cosnier, S.; Labbé, P.; Reverdy, G. A biosensor as warning device for the detection of cyanide, chlorophenols, atrazine and carbamate pesticides. *Anal. Chim. Acta,* **1995**, *311*(3), 255-263.
[http://dx.doi.org/10.1016/0003-2670(94)00686-G]

[304] Touloupakis, E.; Giannoudi, L.; Piletsky, S.A.; Guzzella, L.; Pozzoni, F.; Giardi, M.T. A multi-biosensor based on immobilized Photosystem II on screen-printed electrodes for the detection of herbicides in river water. *Biosens. Bioelectron.,* **2005**, *20*(10), 1984-1992.
[http://dx.doi.org/10.1016/j.bios.2004.08.035] [PMID: 15741067]

[305] Abdalla, N.; Youssef, M.; Algarni, H.; Awwad, N.; Kamel, A. All solid-state poly (vinyl chloride) membrane potentiometric sensor integrated with nano-beads imprinted polymers for sensitive and rapid detection of bispyribac herbicide as organic pollutant. *Molecules,* **2019**, *24*(4), 712.
[http://dx.doi.org/10.3390/molecules24040712] [PMID: 30781449]

[306] Koblížek, M.; Malý, J.; Masojídek, J.; Komenda, J.; Kučera, T.; Giardi, M.T.; Mattoo, A.K.; Pilloton, R. A biosensor for the detection of triazine and phenylurea herbicides designed using Photosystem II coupled to a screen-printed electrode. *Biotechnol. Bioeng.,* **2002**, *78*(1), 110-116.
[http://dx.doi.org/10.1002/bit.10190] [PMID: 11857287]

[307] Shams, N.; Lim, H.N.; Hajian, R.; Yusof, N.A.; Abdullah, J.; Sulaiman, Y.; Ibrahim, I.; Huang, N.M.; Pandikumar, A. A promising electrochemical sensor based on Au nanoparticles decorated reduced graphene oxide for selective detection of herbicide diuron in natural waters. *J. Appl. Electrochem.,* **2016**, *46*(6), 655-666.
[http://dx.doi.org/10.1007/s10800-016-0950-4]

[308] Tripathi, A.; Sundaram, S.; Tripathi, B.C.; Rahman, A. Activity and stability of Herbicide Treated Cyanobacteria as Potential Biomaterial for Biosensors. *Res. J. Environ. Sci.,* **2005**, *5*, 479-485.

[309] Haddaoui, M; Raouafi, N Chlortoluron-induced enzymatic activity inhibition in tyrosinase/ZnO NPs/SPCE biosensor for the detection of ppb levels of herbicide. *Sens Acutators b ,* **2015**, *291*, 171-178.

[310] Mani, V.; Devasenathipathy, R.; Chen, S.M.; Wu, T.Y.; Kohilarani, K. High-performance electrochemical amperometric sensors for the sensitive determination of phenyl urea herbicides diuron and fenuron. *Ionics,* **2015**, *21*(9), 2675-2683.
[http://dx.doi.org/10.1007/s11581-015-1459-2]

[311] Tripathi, A.; Sundaram, S.; Tripathi, B.C.; Rahman, A. Activity and stability of Herbicide Treated Cyanobacteria as Potential Biomaterial for Biosensors. *Res. J. Environ. Sci.,* **2005**, *5*, 479-485.

[312] Songa, E.A.; Arotiba, O.A.; Owino, J.H.O.; Jahed, N.; Baker, P.G.L.; Iwuoha, E.I. Electrochemical detection of glyphosate herbicide using horseradish peroxidase immobilized on sulfonated polymer matrix. *Bioelectrochemistry,* **2009**, *75*(2), 117-123.
[http://dx.doi.org/10.1016/j.bioelechem.2009.02.007] [PMID: 19336272]

[313] Pandard, P; Rawson, D M An Amperometric Algal Biosensor for Herbicide Detection Employing a Carbon Cathode Oxygen Electrode. *Environmental Toxicology and Water Quality,* **1993**, *8*, 323-333.
[http://dx.doi.org/10.1002/tox.2530080309]

CHAPTER 4

Nano Porous Anodic Aluminum Oxide: An Overview on its Fabrication and Potential Applications

Ujjal Kumar Sur[1,*]

[1] *Department of Chemistry, Behala College, University of Calcutta, Kolkata, India*

Abstract: The quick development in nanotechnology has raised the status of this modern technology owing to the decrease in the sizes of structures and devices. There has been considerable dispute concerning the future consequences of nanotechnology. Nanotechnology has the probable ability to design many new and novel materials and devices in smaller dimensions with broad-range applications in medicine, electronics and sustainable energy production. Nano porous anodic aluminum oxide (AAO) films consisting of self-organized hexagonal arrays of invariable parallel nanochannels have been widely applied as the building block to fabricate various functional nanostructures of different morphologies such as nanoparticles, nanowires and nanotubes. These functional nanostructures can be potentially utilized in various applications like magnetic storage media, optoelectronics, bio/chemical sensors, photonics and plasmonics. This chapter describes the different fabrication processes of AAO films in detail along with citation of a few interesting applications.

Keywords: Anodic aluminum oxide, Aluminum, Anodization, Electrochemical replication, Hard anodization, Hexagonal, Nanotechnology, Nanostructures, Nano porous, Nanochannels, Nanoimprint lithography, Plasmonics, Self-organization, Surface-enhanced Raman scattering, SERS substrates.

INTRODUCTION TO NANOTECHNOLOGY

Study of structures of dimensions of 100 nanometers or less (1 nm = 10^{-9} m) and development of new materials or devices with dimensions on this scale is known as nanotechnology. The idea of "nanotechnology" was first proposed by famous American physicist Richard Feynman in his famous lecture titled 'There is plenty of room at the bottom' delivered at the American Physical Society meeting at Caltech in 1959.

---

[*] **Corresponding author Ujjal Kumar Sur:** Department of Chemistry, Behala College, University of Calcutta, Kolkata, India; E-mail: uksur99@yahoo.co.in

Sibel A. Ozkan (Ed.)

Feynman first introduced the concept of nanotechnology and predicted a procedure by which manipulation of individual atoms and molecules can be carried out by means of accurate tools. Nanotechnology started its venture in the early 1980's with two topmost advances; the study of cluster science and discovery of Scanning tunneling microscope by Binnig and Rohrer.

Nanotechnology got a major boost when integrated circuits (ICs) were invented in 1958. According to Moore's law [1], the number of transistors in the computing hardware which can be located economically on an integrated circuit has augmented exponentially, becoming doubled roughly every two years. Dr. Gordon E. Moore, the co-founder of Intel, first observed this in 1965 (Fig. **1**).

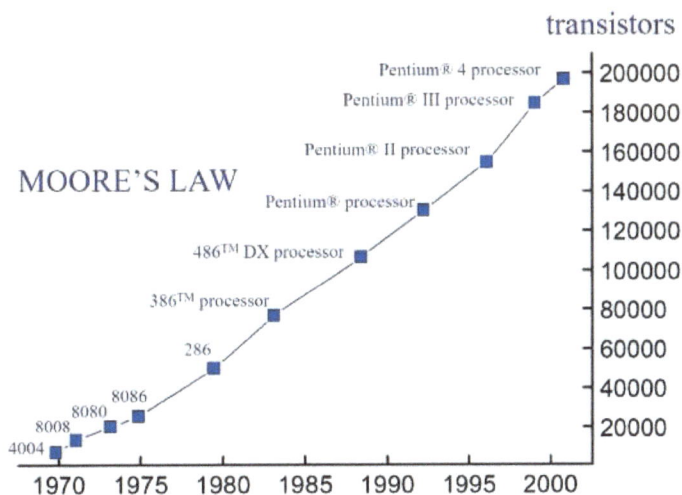

Fig. (1). Schematic diagram showing Moore's law.

As the present computing-hardware technology has fizzled out in 2012 regarding the fabrication procedure and device operation. Consequently, alternate patterning techniques and computation outlines are required. Introduction of quantum, molecular and optical computers, carbon-nanotubes based devices can provide a new dimension of computation. Carbon based nanomaterials such as fullerenes, carbon nanotubes and graphene had transformed the computing-hardware technology and one can expect that these carbon nanomaterials based devices will substitute the prevailing silicon (Si)-technology based devices in the near future by providing an overall upgradation of computation speed and efficiency.

Different physical properties like mechanical, optical, electrical, *etc.* can be modified meaningfully by reducing the dimension of the system and size-related intensive properties, such as quantum confinement in semiconductor nanomaterials, surface plasmon resonance in metal nanoparticles will be exhibited by nanostructured materials. The mechanical, thermal and catalytic properties of these nanomaterials will change in comparison to the bulk materials due to the increase in surface area to volume ratio.

WHY DO WE WANT TO FABRICATE NANOSTRUCTURES ?

In addition to fundamental physical interest in the nanometer size regime, the different physical properties of nanosized structures compared to their bulk and molecular counterparts enable unique potential technological applications in medicine, electronics and optical devices.

INTRODUCTION TO POROUS ANODIC ALUMINUM OXIDE

Anodic aluminum oxide (AAO) films can be grown employing the electrochemical oxidation of an aluminum (Al) substrate in an electrolytic cell (Fig. **2**). AAO films have been studied and applied in different industrial applications, including the construction of electrically insulating layers (dielectrics), anti-corrosive coatings and decorative coloration of metal surfaces [2, 3]. This fabrication procedure of AAO films is commonly known as anodization. Nano porous AAO films with hexagonal arrangement of monodisperse nanopores are common template systems for the fabrication of various functional nanostructures (*e.g.* nanofabrication of quantum dots) [4 - 6]. In 1995, self-organized AAO film was discovered by Japanese scientists Masuda and Fukuda using the right anodization conditions [7].

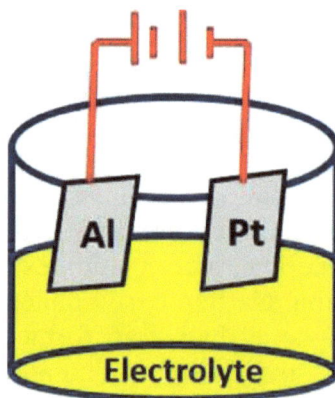

Fig. (2). Typical electrolytic cell used for anodization of Al.

Conventionally, a two-step anodization process is employed to produce such self-organized arrays of AAO nanochannels. Some of the applications of porous AAO films are mentioned as follows:

- Exploration of the unique optical, electrical and magnetic quantum confinement of porous AAO films for fundamental research.
- Microfiltration (In gas filtrations and biological separations).
- Template for fabrication of optical waveguides and photonic crystals for optical circuits.
- Template for fabrication of ordered arrays of quantum dots for lasers and photodetectors.
- Template for carbon-nanotube growth for electronic, mechanical applications.
- Ultra-large scale integration (ULSI) memory devices and integrated circuits (IC).

HISTORICAL TIMELINE OF POROUS AAO FILMS

Anodization was first introduced at the industrial scale in 1923. Since then, several research groups had fabricated anodic aluminum oxide films by electrochemical oxidation of aluminum [8 - 10].

1. 1920's - Porous AAO films had started to be used commercially to protect and finish bulk Al surfaces.
2. 1940-1960's - The first characterization of structure of porous AAO with the arrival of electron microscopy.
3. 1950 - Keller *et al.* first reported the mechanism of self-organization of alumina [11].
4. 1970 - Manchester research group carried out the first real experimental work and demonstrated the dependency of pore diameter on the anodization potential, *etc.*
5. 1970 - O'Sullivan *et al.* [12] first reported the morphology mechanism of formation of porous anodic films on aluminum.
6. 1992 - V. P. Parkhutik and V. I. Shershulsky carried out the first quantitative theoretical attempt to explain the pore growth from first principles [13].
7. 1995 - Japanese group (Masuda *et al.*) discovered the right anodization conditions under which pores in AAO films will self-organize into close packed arrays [7].
8. 1996- present - Use of porous AAO films for nano-applications.
9. 1998 - Although the mechanism of self-ordering in porous AAO films is still not clear, German scientists from Max-Planck Institute of Microstructure Physics proposed one possible self-organization mechanism [14].

OVERVIEW OF ANODIC ALUMINUM OXIDE FILMS

Dissimilar morphology of the grown anodic aluminum oxide can be monitored relying on the chemical nature of the electrolytic medium. Acidity of electrolyte can be regarded as the main influencing aspect to the different growth behaviour. Two types of AAO films can be produced employing anodization process by changing the pH or acidity of electrolyte as shown in Fig. (3).

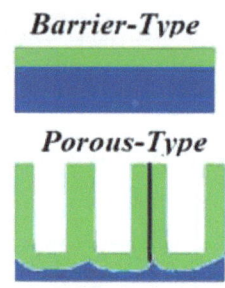

Fig. (3). Diagram showing the porous and barrier type anodic aluminum oxide films.

Porous type AAO film:

- Grown oxide slightly soluble in electrolyte.
- Aqueous sulfuric (H_2SO_4), oxalic acid ($H_2C_2O_4$) and phosphoric (H_3PO_4) acids are used as electrolyte during anodization. Other acids such as selenic, tartronic, malonic, phosphonic, tartaric, phosphonoacetic, etidronic, malic, and citric acids can be employed for the fabrication of porous AAO by anodization. Nishinaga *et al*. [15] had employed selenic acid to fabricate porous AAO films, which possessed great potential of low porosity, colorlessness, and high transparency. It also does not show photoluminescence as compared to the oxide formed with organic electrolyte.

Barrier type film:

- Grown oxide insoluble in electrolyte. Anodization in a neutral electrolyte such as borate, oxalate, citrate *etc*.;
- Nearly neutral (pH = 5-7) electrolyte is used during anodization.

POROUS AAO FILMS

- Anodization of Al in acidic electrolyte (*e.g.* oxalic acid).
- Oxide will produce at the metal/oxide and oxide/electrolyte interfaces. Pores initiation will commence at random positions due to field assisted dissolution at the oxide/electrolyte interface.
- Appropriate potentials and long anodization times are needed for ordering.
- Repulsion between adjoining pores will cause ordering, which can arise as a result of the mechanical stress at the metal/oxide interface.

OUTLINE OF ANODIZATION MECHANISM

Two scientists from Russia, V. P. Parkhutik and V. I. Shershulsky first framed a numerical model to elucidate the mechanism of pore-growth throughout the electrochemical anodization of Al from first principles [13]. Fig. (**4**) shows the schematic diagram of the kinetics of porous oxide growth during the anodization of Al. The various steps of porous structure development are also demonstrated in the diagram. Oxide growth will progress through ionic conduction and reaction of Al^{3+} and oxygen containing anions by applying an electric field (*e.g.* 2 Al^{3+} + 3 OH^- → Al_2O_3 + $3H^+$ + $6e^-$).

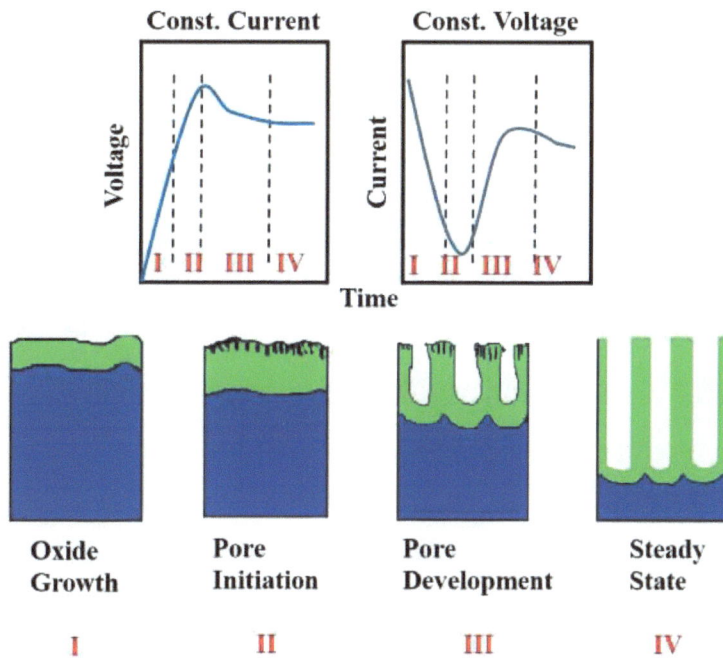

Fig. (4). Schematic diagram showing the kinetic of porous oxide growth during the anodization of Al in galvanostatic and potentiostatic regimes. The stages of porous structure development are also shown. (Reprinted with permission from reference 8 [V. P. Parkhutik and V. I. Shershulsky, *J.Phys.D: Appl. Phys.*, 1992, vol. **25**, 1258]. Copyright [1992], Institute of Physics).

- Pores will initiate at random positions through the dissolution of the oxide in presence of electric field at the oxide/electrolyte interface.
- Initially, oxide growth dominates (stage I). Later, dissolution of oxide film becomes competitive. Barrier layer becomes thinner and pores initiate (stage II).
- When both the oxide growth and dissolution mechanisms occur roughly at the same rate (stage III and IV), steady state approaches.

BARRIER TYPE AAO FILM

- Oxide growth proceeds at the Al anode (+).

- H_2 gas is evolved at the Pt cathode (-).

- The current between the cathode and anode is carried by the

electrolyte.

- Oxidation reaction occurs at the Al anode:

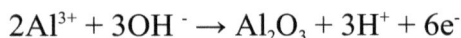

$2Al^{3+} + 3OH^- \rightarrow Al_2O_3 + 3H^+ + 6e^-$

- Electrolysis of water at the aluminum oxide/electrolyte interface.

- Reduction reaction occurs at the Pt cathode:

$2H^+ + 2e^- \rightarrow H_2$

The overall electrochemical reaction is shown below.

$2Al + 3 H_2O \rightarrow Al_2O_3 + H_2$

Fig. (**2**) shows a typical electrolytic cell used for anodization of Al.

ORDERED GROWTH OF POROUS AAO FILM

In 1995, Masuda and Fukuda from Japan found that pores will self-organize under right anodization conditions [7]. The two most important conditions are narrow voltage ranges and long anodization time. The long-range organization has been detected to take place under restricted voltage conditions which are explicit to the electrolyte solution applied for anodization, that is long-range organization takes place at 25, 40 and 195 V respectively for H_2SO_4, $H_2C_2O_4$ and H_3PO_4 solutions corresponding to interpore distance of 63, 100 and 500 nm, respectively [16 - 18]. Polycrystalline structure with perfectly ordered domains of a few microns in size are obtained in these three acidic electrolytes for long anodization time and appropriate voltages with defects occurring at the grain boundaries.

Fig. (**5**) shows the schematic picture of porous AAO film structure. The oxide film takes a cellular structure with a central pore in each cell. The schematic illustrates undeviating hexagonal cells, but utmost anodization conditions will yield films with supplementary disorder, with a distribution of cell size and pore diameter. The dimension of cell is in the range of 50-300 nm having pore diameter classically 1/3 to 1/2 of the cell size. The cell population density is roughly from 10 to 100 per μm^2 with aspect ratio of ordered nanostructure of the order of 1000: 1. For instance, AAO films typically grown in sulfuric acid have film thickness of 20 to 50 μm with 20 nm pores.

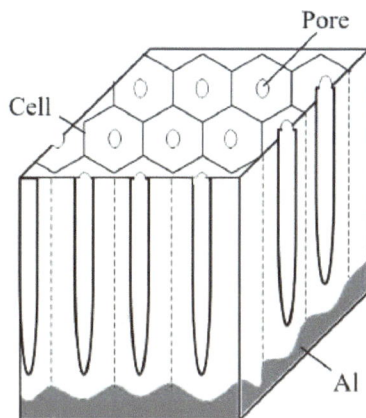

Fig. (5). Schematic diagram of an idealized structure of porous AAO film. (Reprinted with permission from reference 16, copyright [2001], The Electrochemical society).

Li *et al.* observed the self-organization of two-dimensional pore arrays with interpore distances of 50-420 nm in porous AAO films during the growth of film in H_2SO_4, $H_2C_2O_4$ and H_3PO_4 solutions [14]. Fig. (**6**) illustrates the schematic diagram of an idealized structure of porous AAO film. Fig. (**7**) shows the plot of interpore distance *versus* anodization voltage in self-organized porous AAO for H_2SO_4, $H_2C_2O_4$ and H_3PO_4 solutions. The solid line represents the relation d = -1.7 + 2.81 U_a [19]. Fig. (**8**) illustrates the scanning electron microscopy (SEM) images of the bottom view of porous AAO layers grown in H_2SO_4, $H_2C_2O_4$ and H_3PO_4 solutions.

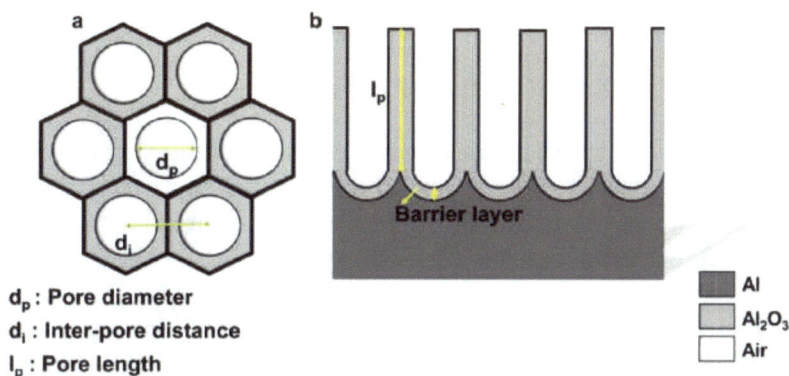

d_p : Pore diameter
d_i : Inter-pore distance
l_p : Pore length

Fig. (6). Schematic diagram of an idealized structure of porous AAO film. (Reprinted with permission from reference 75, copyright [2017], Wiley).

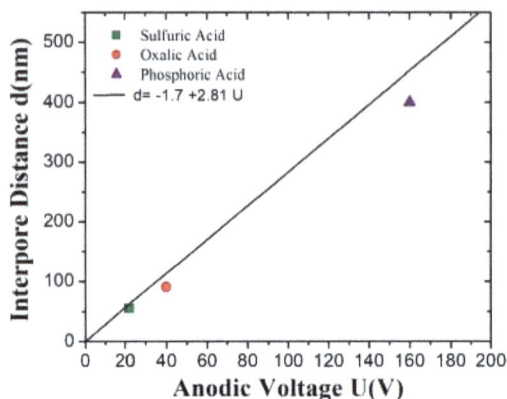

Fig. (7). Plot of interpore distance (d) *versus* anodization voltage (U_a) in self-organized porous AAO film for H_2SO_4, $H_2C_2O_4$ and H_3PO_4 solutions. The solid line represents the relation d = -1.7 + 2.81 U_a (after reference 13). (Reprinted with permission from reference 9 [A. P. Li, F. Muller, A. Birner, K. Nielsch and U. Gosele, *J. Appl. Phys.*, 1998, vol. **84**, 6023]. Copyright [1998], American Institute of Physics).

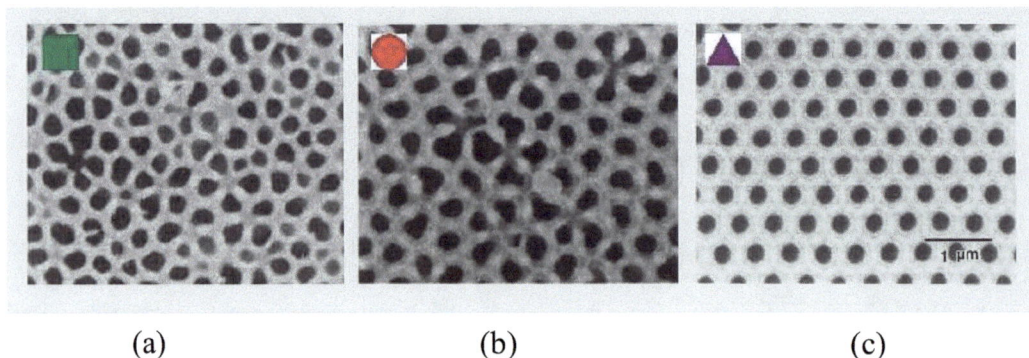

(a)   (b)   (c)

Fig. (8). SEM images of the bottom view of porous AAO films. **(a)** anodization in 0.3 M sulphuric acid at 25 V; **(b)** anodization in 0.3 M oxalic acid at 40 V; **(c)** anodization in 10% (by weight) H_3PO_4 at 160 V. (Reprinted with permission from reference 9 [A. P. Li, F. Muller, A. Birner, K. Nielsch and U. Gosele, *J. Appl. Phys.*, 1998, vol. **84**, 6023]. Copyright [1998], American Institute of Physics).

TYPES OF ANODIZATION PROCESS

Two types of anodization processes are generally employed in the Al industry for the development of thick AAO films. One is mild anodization while the other is hard anodization. Hard anodization process was introduced in the early1960's [20]. Hard anodization can be carried out by high voltage applying H_2SO_4 as electrolytic solution, which causes the fast formation of thick porous oxide layer. These thick porous oxide layers are more chaotic than those formed by mild anodization. However hard anodization procedure has not been carried out in academic research and employed to fabricate nanomaterials until now owing to the complications in regulating significant basic parameters like pore size, interpore distance and the aspect ratio of porous AAO films.

Woo Lee and his group from the Max-Planck Institute, Germany had carried out a novel, high-speed anodization process based on oxalic acid for the fast fabrication of self-organized, long-range ordered AAO films [21]. This is a new type of the supposedly hard anodization method which has been extensively applied industrially from 1960's for several industrial applications like surface finishing of Al cookware, automobile engineering, textile equipment *etc*. By contrast, mild anodization process employing Masuda's protocol has not been applied industrially owing to the slow oxide growth rate (2-5 μmh^{-1}). A simple and rapid production of highly ordered AAO with a wide range of pore sizes and interpore distances would be extremely needed practically.

Highly ordered porous AAO template had been fabricated by Lee and his co-workers by pre-anodization of the Al surface, judicious control of the surface-heat evolution and suitable choice of the anodization voltages [21]. These AAO films were thicker and were fabricated at a rate 25-35 times faster than those produced by mild anodization. They proposed that self-controlling cell development in presence of high current density at the lowermost portion of the pores is also related to the mechanical stress at the metal/oxide interface, which is primarily accountable for the growth of highly well-organized porous structures during hard anodization. The researchers understood from their experimental results about the structures of well-organized AAO membranes with high aspect ratios (> 1000), persistent and sporadically controlled pore sizes. These well-defined 3-dimensional (3-D) nanoarchitectures can provide a wide-range of nanotechnological applications like 3-D photonic crystals, meta-materials, microfluidics and template based synthesis of multifunctional nanowires and nanotubes.

FABRIATION OF SELF-ORGANIZED ARRAYS OF POROUS AAO NANOCHANNELS

Self-organized arrays of porous AAO nanochannels [7] can be grown by employing a two-step anodization process based on Masuda's work in 1995. Al sample can be electrochemically polished to nanometric smoothness prior to anodization to fabricate these self-organized AAO films. Highly pure annealed Al can be electrochemically polished for approximately half an hour in a mixture of $HClO_4$ and C_2H_5OH (ratio 1:5) at 5°C under constant stirring. The root mean square (rms) roughness of the Al surface turns out to be less than a few nanometers after electrochemical polishing. The smooth Al foil obtained after electrochemical polishing can be anodized for 5 h in 0.3 M oxalic acid at 3° C under the influence of constant anodization voltage of 40 V. The AAO film can be dissolved in 3% (by weight) CrO_3 solution after the first anodization step. The self-organized arrays of AAO nanochannels of anticipated thickness can be obtained by the second anodization step for long anodization time. Lastly, pore widening of the nanochannels can be achieved employing 5% H_3PO_4 (by weight) solution. Fig. (9) displays the SEM image of an array of self-organized AAO nanochannels. Biring *et al.* fabricated large-scale ultra-smooth Al foils using fast electrochemical replication (ER) technique and subsequently, used these Al foils as substrates to produce self-organized and long-range ordered arrays of AAO nanochannels [22] employing a novel protocol. In this procedure, widely accessible large-scale smooth Si wafers were used as master substrates and Al was electrochemically deposited onto it. The replicated Al foil can be easily detached from the master Si surface without damaging either the replicated Al or the Si master surfaces owing to the tunability of the adhesion of the electrochemically deposited Al on Si. The replicated Al foil had an rms roughness of ~ 1 nm analogous to that of Si master. Fig. (10) illustrates the schematic diagram exhibiting the ER method to fabricate ultra-smooth Al foil. The potentially explosive electrochemical polishing step can be prevented by employing the ER method to fabricate ultra-smooth Al foils.

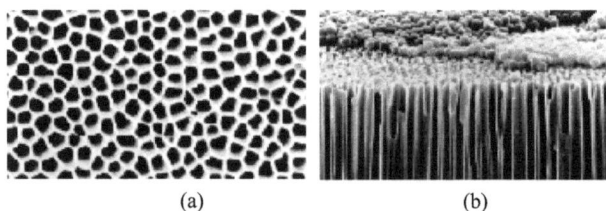

(a)   (b)

Fig. (9). SEM image of an array of self-organized AAO nanochannels. **(a)** Top view, **(b)** Cross-sectional view.

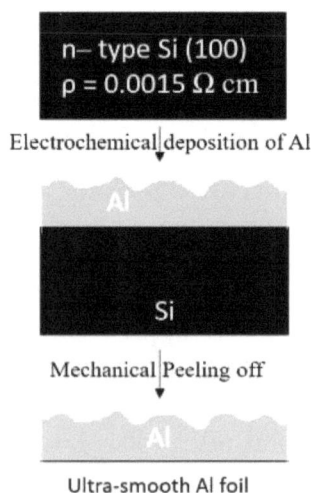

Fig. (10). Schematic diagram showing the electrochemical replication technique to fabricate ultra-smooth Al foil. (Reprinted with permission from reference 15 [S. Biring, K. T. Tsai, U. K. Sur and Y. L. Wang, *Nanotechnology*, 2008, vol. **19**, 015304]. Copyright [2008], Institute of Physics).

FABRICATION OF ARRAYS OF LONG-RANGE ORDERED AAO NANOCHANNELS

Highly ordered arrays of AAO nanochannels over large area (> 100 μm^2) are important to fabricate various devices of nanometer dimensions such as electronics, optoelectronics and micro electromechanical devices. Various lithographically guided techniques have been employed to fabricate such highly ordered arrays of AAO nanochannels.

Masuda and his co-workers used nanoimprint lithography technique to grow defect free array of long range ordered AAO nanochannels [23]. They used silicon carbide (SiC) mold with hexagonally ordered arrays of convexes fabricated by electron-beam (EB) lithography. Then, using this SiC mold, patterned Al substrates with hexagonal arrays of nano concaves were obtained. Subsequently, these patterned Al substrates were anodized in oxalic acid solution under proper anodization conditions to grow long range ordered AAO nanochannels. Nano concaves present on the patterned Al substrates can behave as the starting point and can control the evolution of nanochannels arrays organized in two-dimensional hexagonal configurations. Fig. (11) demonstrates the schematic diagram for the development of highly ordered AAO nanochannel arrays by nanoimprint lithography technique employing SiC mold.

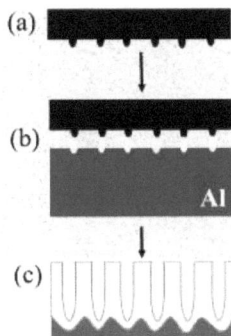

Fig. (11). Schematic diagram for the growth of highly ordered AAO nanochannels arrays by nanoimprint lithography using SiC mold. (Reprinted with permission from reference 16 [H. Asoh, K. Nishio, M. Nakao, T. Tamamura and H. Masuda, *J. Electrochem. Soc.,* 2001, vol. **148**, B152]. Copyright [2001], The Electrochemical Society).

Ordered AAO nanochannels were produced by Liu *et al.* on focused-ion-beam (FIB) lithographically prepatterned Al surfaces [24]. Arrays of hexagonally closed packed concaves were created on electrochemically polished polycrystalline Al substrates by employing a commercial 50 keV Ga FIB lithography. In the consequent one-step anodization process, the growth of well-organized nanochannels were guided by the concaves. Highest organized areas of nanochannels with average interpore distance of 100 nm were grown by employing 40 V anodization voltage in 0.3 M oxalic acid solution at 3° C. Liu *et al.* fabricated 10 μm thick ordered arrays of AAO films after a few hours of anodization after selecting the same lattice parameter for the arrays to increase the prospect of guided growth.

A fast electrochemical replication technique was employed by Biring *et al.* to fabricate patterned Al nanostructures [25]. The replicated nanostructured Al films with hexagonally close-packed guiding nano-concaves were employed to subsequently produce long range ordered arrays of AAO nanochannels in a single-step anodization process. Conventionally, nanoimprint lithography can be applied to fabricate patterned Al substrates, which can be later anodized for the mass production of long-range ordered AAO nanochannels [26, 27].

Fig. (**12**) displays the schematic diagram explaining the electrochemical replication process for the fabrication of nanostructured Al film. Al can be electrochemically deposited onto a lithographically patterned Si master using a non-aqueous organic hydride bath of aluminum chloride and lithium aluminum hydride at room temperature in diethyl ether in this high fidelity replication technique. Si master was chemically pretreated to permit an effortless detachment of the replicated Al foil from the master, allowing its tedious usage for mass replication. Patterning onto metal substrates like Al by applying nanoimprint lithographic technique involves high pressures and mutilation of the imprint master frequently takes place after numerous usage pressure to fabricate patterned Al substrate just like nanoimprint lithographic techniques. The technique can be further employed for the development of highly ordered AAO nanochannels over large areas as the dimension of the nanopatterned Si master can be easily enlarged using the present Si lithographic technology.

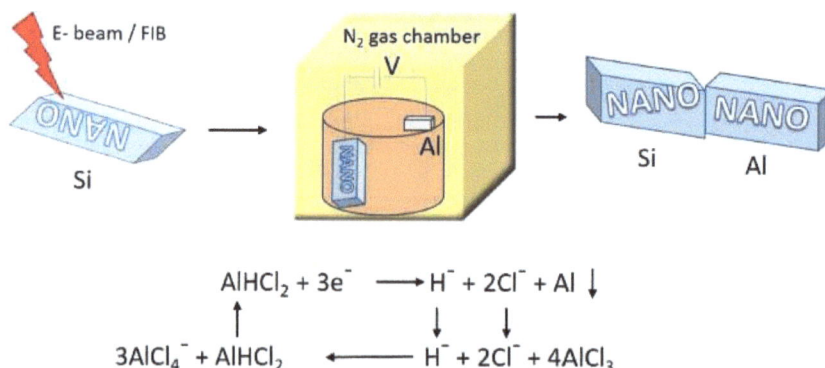

$$AlHCl_2 + 3e^- \longrightarrow H^- + 2Cl^- + Al \downarrow$$

$$3AlCl_4^- + AlHCl_2 \longleftarrow H^- + 2Cl^- + 4AlCl_3$$

Fig. (12).   Schematic diagram showing the electrochemical replication process for the fabrication of nanostructured Al foil.

Applications of Porous AAO Films

Nanoporous anodic alumina can be utilized as precursors for templates, photonic structures, membranes, drug delivery platforms or nanoparticles. Other than the use of aluminum oxide films as filtration membranes, they are regularly used to produce high aspect ratio nanowires [28]. Various materials, like metals, both magnetic [29] and nonmagnetic [30], semiconductors [31] and even heterostructures [32], had been grown in the porous membranes employing mainly electrochemical deposition techniques. Besides, porous aluminum oxide membranes have also been employed as photonic crystals [33], as humidity sensors [34], or as cathodes for organic light emitting diodes [35]. Nanoporous anodic aluminium oxide has been widely used for the development of various functional nanostructures. There are many fields in which nanoporous anodic alumina structures have demonstrated their utility. Table **1** illustrates some of the recently established potential applications of porous AAO nanostructures. Porous AAO nanostructures based template-synthesized multi-component nanowire systems can be used as sensors and actuators. Sensors and actuators are among the most important components in many electronics devices and products. Multi-component nanowire based sensors and actuators have potential applications in a broad variety of mechanical, electrical, thermal, and biomedical devices as well as in microrobotics. Nanowires are one of the most promising nano-building blocks for many applications, including sensors/biosensors, electronics, photonics, energy conversion and storage devices, MEMS and NEMS, biotechnology and nanomedicine, microrobotics as well as nanorobotics. Table **1** illustrates some of the recent applications of porous AAO nanostructures.

Table 1. Recent applications of nano porous AAO-based structures.

Application	Reported Utility	References
Photonic structures	2-D photonic crystals in the visible wavelength	[36]
-	Photonic bandgap sensor	[39]
Sensors	Chemical and biosensors	[55]
-	Detection of lead ions	[59]
-	Quantitative detection of trypsin	[60]
-	Salmonella sensing	[61]
-	Humidity Sensor	[34]
Membranes for filtration and separation	Filtering and Separation of water from bacteria	[64, 65]
Diagnostics and pathogen detection	Rapid identification of pathogens	[56]
Template	High aspect ratio nanowire	[28]

(Table 1) cont.....

Application	Reported Utility	References
-	Fabrication of terabit magnetic storage devices	[75]
-	Cathodes for organic light emitting diodes	[35]
Biological monitoring and cell culture	mimic and observe biological systems at the cellular level	[66, 67]
Drug delivery	Controlled drug delivery both *in vitro* and *ex vivo* and the release of drug inside the bone	[69]
Microrobotics	targeted drug delivery, environmental control, biopsy and surgery based on microrobotics	[71 - 73]

Photonic Crystal Using Porous AAO

The optical properties of two-dimensional materials like photonic crystals with a spatially periodic index of refraction have potential applications to chalk out novel optoelectronic devices. Nevertheless, it is extremely hard to produce an exactly ordered periodic structure with anticipated aspect ratios. There are very few reports on the fabrication of 2-D photonic crystals in the near infra-red (IR) or visible wavelength region owing to this reason. Masuda and his co-workers had fabricated 2-D photonic crystals in the visible wavelength region employing highly ordered porous AAO as template material [36]. Photonic band-gap structures based on porous AAO can be conveniently fabricated with easily controllable geometry. Additionally, porous AAO is appropriate to produce optical waveguide. Masuda and his group measured the transmission properties of porous AAO with a lattice constant from 200 to 250 nm in the visible wavelength section for discrete propagation directions.

Nanoporous anodic alumina can be utilized as a photonic crystal (PC)—typically through sinusoidal anodization method which has been described in a few review articles [37, 38]. PCs are structures with a sporadic variation of refractive index, interweaving areas with high and low refractivity. As an effect, some wavelengths can propagate in the structure while others cannot—these are described as the photonic band gaps, which can be utilized as optical sensors. A typical application of such sensor is a real-time monitoring of the interaction between human proteins and heavy ions [39].

Production of Gold Nanodot Arrays Employing Porous AAO as an Evaporation Mask

Fabrication of nanometer scale metal dot array had fascinated significant attention owing to the development of various electronic and optoelectronic devices. Generally, EB lithography is employed to fabricate these structures. However, the nanofabrication by EB lithography has several disadvantages like low output due to the lengthy contact time and high cost of electron-beam instrument.

Masuda and Satoh had fabricated a highly ordered array of gold nanodots on Si substrate by vacuum evaporation mask [17]. The technique demonstrated in their report is suitable to fabricate nanodot arrays over a large area employing a broad variety of materials that can be deposited by vacuum evaporation. Nanodot arrays obtained by this process will be further useful in the fabrication process for nanometer scale semiconductor devices.

Fabrication of Ag/AAO System in the Form of Bio/Chemical Sensors

Surface-enhanced Raman scattering (SERS) has developed into a versatile and popular surface sensitive spectroscopic analytical tool owing to the gigantic augmentation of weak Raman signal in the vicinity of plasmonic nanomaterials and can provide appropriate detection of various chemical and biological systems. SERS technique has diverse applications ranging from plasmonics, sensing, catalysis to biomedical applications and diagnostics. Fabrication of SERS active substrates is challenging and has an important role in controlling the SERS activity.

Various plasmonic nanomaterials can be employed as SERS active substrates in SERS technique and SERS enhancement factor will regulate their efficiency as a SERS substrate [40 - 43]. In SERS-based sensing applications, the optical trans-ducer or the SERS substrate and the molecular functional interface are the two primary prerequisites that will regulate the overall development of sensors. The optical properties of a SERS substrate will affect the analytical sensitivity, whereas the functional interface of a SERS substrate will facilitate the detection specificity. A great deal of work has been done in designing different types of plasmonic nanomaterial to increase the SERS enhancement factor, henceforth the analytical sensitivities. Spherical gold or silver nanoparticles or substrates [44 - 47] can act as the first generation of SERS substrates. Later, anisotropic nanoparticles such as rod, cubic, triangular and star-shaped plasmonic nanomaterials and two-dimensional plasmonic substrates were employed [48 - 50]. Higher SERS enhancement factors could be accomplished [50] owing to the higher electromagnetic field conveyed by the antenna effect of the strong features of the second-generation SERS substrates. Various nanoparticle assemblies and three-dimensional nanoscale plasmonic substrates were developed next as the next generation of SERS substrates containing plasmonic "hot spots", where the SERS enhancement factor is much higher [51, 52]. In these nanoparticle assemblies, Raman active molecules are chemically or physically positioned at the slim gap between nanoparticles ("SERS hot spot"), where they can generate the maximum SERS enhancement [53, 54]. The development of SERS substrates has given rise to 10-12 orders of maximum Raman signal enhancement, which could facilitate even single-molecule detection under optimum conditions. Generally, SERS-based sensing methods have become more popular and beneficial over traditional spectroscopic methods like fluorescence as a result of ultra-sensitivity even up to single-molecule detection level.

Arrays of silver (Ag) nanoparticles with a precisely controlled gap up to 5 nm were grown electrochemically by Wang *et al.* employing porous AAO film as template [55]. The fabricated Ag/AAO system with tunable sub-10 nm interparticle gap was utilized as an effective surface-enhanced Raman Scattering (SERS) active substrate with a large enhancement factor ($\sim 10^8$). From their experimental results, it was concluded that the "hot junction" existing at the interparticle gap of this nanostructure-based SERS substrates will enhance the SERS sensitivity, which is a major aspect for huge electromagnetic field enhancement and augmented, highly uniform SERS signal. This Ag/AAO based SERS substrate was extremely uniform and can provide reproducible SERS signals. Therefore, the fabricated SERS substrates can be used as bio/chemical sensor to detect concentration up to picomolar level. This Ag/AAO based SERS substrate was further used for rapid identification of pathogens [56]. Therefore, this Ag/AAO based SERS substrates are novel and effective, which can be further applied into single molecular level, enhancing both the detection limit and SERS sensitivity.

Sensors

In recent years, the unique geometry of porous AAO films has been demonstrated in many fields, for example, as sensing tools [57], providing a promising alternative over plasmonic nanoparticles. Porous AAO films were utilized as a shadow mask to fabricate a surface covered with plasmonic dimers as reported by Schmidt *et al.* [58]. Tabrizi *et al.* demonstrated the versatility of AAO nano porous structures through the accurate detection of specific biomolecules and ions. Sensing applications such as determination of lead (II) ions [59], quantitative detection of trypsin [60], or Salmonella sensing through the recognition of specific DNA fragments [61] were cited in the literature.

Membranes for Filtering and Separation

The nanostructured array of pores of porous AAO films is extremely consistent. Layers up to several micrometers are essential to fabricate self-standing and sturdy membranes [62] and the inner side of their walls can be exactly designed for specific application [63]. Stroe *et al.* [64] employed Porous AAO membranes with zinc oxide nanosheets grown on the surface and the pore size enabled to separate bacteria from water [65]. It was demonstrated that the E. coli population will reduce by 73% in UV light over 24 h.

Biological Monitoring and Cell Culture

The dimension of porous AAO structures permits unique possibility to mimic and observe biological systems at the cellular level [66]. Chen *et al.* employed porous AAO films coated with CuO and L-Cys/D-Cys to monitor the response of biomolecules [67]. Moreover, the macrophage response can be modified with varying the pore size [68].

Drug Delivery

Cost-effective, biocompatible structures of nanoporous AAO films along with derived materials can be engineered in all dimensions and can be further employed for drug delivery. One such example can be an implant based on the aluminum wire surface-modified through anodization. Thin layer of nanoporous AAO films provided on the wire implants was studied for controlled drug delivery both *in vitro* and *ex vivo* and the release of drug inside the bone was monitored [69]. The implant provided a stable and sustained release. Viable osteocytes in the implant surroundings detected with the bone histology demonstrated the biocompatibility of such devices.

Microrobotics and Nanorobotics Applications

Chen *et al.* [70] applied FeGa@P(VDF-TrFE) core shell nanowires for nanorobotic applications. Microrobotic systems had demonstrated capable outcomes in numerous applications like targeted drug delivery, environmental control, biopsy and surgery [71, 72]. A nominally invasive surgery can be carried out by means of accurate and controlled micro-manipulation of the microrobots. Vyskocil *et al.* [73] fabricated Au-Ag-Ni microrobotic scalpels with asymmetric bent surface through a sequential electrodeposition of each segments in AAO templates, accompanied by partial etching of Ag segment with H_2O_2.

CONCLUSION

Nano porous AAO films with self-organized and long range ordered arrays have been extensively used as a popular template system to fabricate various functional nanostructures and nanodevices. Various nanodevices had been fabricated so far by utilizing this nano template. For instance, AAO has been utilized as a substrate to fabricate terabit magnetic storage devices [74] as well as chemical/biological sensors [55, 75]. The advent of AAO technology is expected to bring a new dimension to the present computing-hardware technology. Therefore, porous AAO film with various functional nanostructures and nanodevices can be potentially utilized in various nanotechnological applications.

ACKNOWLEDGEMENTS

I would like to thank my co-workers who have contributed to this work. I would like to thank Professor Tapan Ganguly and Professor Yuh-Lin Wang for important discussion. I would like to acknowledge financial support from the projects funded by the DHESTBT, Government of West Bengal (memo no. 161(sanc)/ST/P/S&T/9G-50/2017 dated 8/2/2018).

REFERENCES

[1]   Moore, G.E. Cramming more components onto integrated circuits. *Electronics (Basel),* **1965**, *38*, 1-4.

[2]   Thompson, G.E.; Furneaux, R.C.; Wood, G.C.; Richardson, J.A.; Goode, J.S. Nucleation and growth of porous anodic films on aluminium. *Nature,* **1978**, *272*, 433-435.
[http://dx.doi.org/10.1038/272433a0]

[3]   Diggle, J.W.; Downie, T.C.; Goulding, C.W. Anodic oxide films on aluminum. *Chem. Rev.,* **1969**, *69*, 365-405.
[http://dx.doi.org/10.1021/cr60259a005]

[4]   Lee, W.; Scholz, R.; Nielsch, K.; Gösele, U. A template-based electrochemical method for the synthesis of multisegmented metallic nanotubes. *Angew. Chem. Int. Ed.,* **2005**, *44*(37), 6050-6054.
[http://dx.doi.org/10.1002/anie.200501341] [PMID: 16124018]

[5]   Lee, S.B.; Mitchell, D.T.; Trofin, L.; Nevanen, T.K.; Söderlund, H.; Martin, C.R. Antibody-based bio-nanotube membranes for enantiomeric drug separations. *Science,* **2002**, *296*(5576), 2198-2200.
[http://dx.doi.org/10.1126/science.1071396] [PMID: 12077410]

[6]   Zhi, L.; Wu, J.; Li, J.; Kolb, U.; Müllen, K. Carbonization of disclike molecules in porous alumina membranes: toward carbon nanotubes with controlled graphene-layer orientation. *Angew. Chem. Int. Ed.,* **2005**, *44*(14), 2120-2123.
[http://dx.doi.org/10.1002/anie.200460986] [PMID: 15736234]

[7]   Masuda, H.; Fukuda, K. Ordered metal nanohole arrays made by a two-step replication of honeycomb structures of anodic alumina. *Science,* **1995**, *268*(5216), 1466-1468.
[http://dx.doi.org/10.1126/science.268.5216.1466] [PMID: 17843666]

[8]   Diggle, J.W.; Downie, T.C.; Goulding, C.W. Anodic oxide films on aluminium. *Chem. Rev.,* **1969**, *69*, 365-405.
[http://dx.doi.org/10.1021/cr60259a005]

[9]   Hunter, M.S.; Fowle, P. Determination of barrier layer thickness of anodic oxide coating. *J. Electrochem. Soc.,* **1954**, *101*, 481-485.
[http://dx.doi.org/10.1149/1.2781304]

[10]   Thompson, G.E.; Wood, G.C. Porous anodic film formation on aluminium. *Nature,* **1981**, *290*, 230-232.
[http://dx.doi.org/10.1038/290230a0]

[11]   Keller, F.; Hunter, M.S.; Robinson, D.L. Structural features of oxide coatings on aluminium. *J. Electrochem. Soc.,* **1953**, *100*, 411-419.

[http://dx.doi.org/10.1149/1.2781142]

[12] O'Sullivan, J.P.; Wood, G.C. Morphology and mechanism of formation of porous anodic films on aluminum. *Proc. Roy. Soc. Ser. A Math. Phys. Sci.,* **1970**, *317*, 511-543.

[13] Parkhutik, V.P.; Shershulsky, V.I. Theoretical modelling of porous oxide growth on aluminum. *J. Phys. D Appl. Phys.,* **1992**, *25*, 1258-1263.
[http://dx.doi.org/10.1088/0022-3727/25/8/017]

[14] Li, A.P.; Muller, F.; Birner, A.; Nielsch, K.; Gosele, U. Hexagonal pore arrays with a 50-420 nm interpore distance formed by self-organization in anodic alumina. *J. Appl. Phys.,* **1998**, *84*, 6023-6026.
[http://dx.doi.org/10.1063/1.368911]

[15] Nishinaga, O.; Kikuchi, T.; Natsui, S.; Suzuki, R.O. Rapid fabrication of self-ordered porous alumina with 10-/sub-10-nm-scale nanostructures by selenic acid anodizing. *Sci. Rep.,* **2013**, *3*, 2748.
[http://dx.doi.org/10.1038/srep02748] [PMID: 24067318]

[16] Masuda, H.; Hasegawa, F.; Ono, S. Self-ordering of cell arrangement of anodic porous alumina formed in sulfuric acid solution. *J. Electrochem. Soc.,* **1997**, 144.

[17] Masuda, H.; Satoh, M. Fabrication of gold nanodot array using anodic porous alumina as an evaporation mask. *Jpn. J. Appl. Phys,* **1996**, *35*, 126-129.
[http://dx.doi.org/10.1143/JJAP.35.L126]

[18] Masuda, H.; Yada, K.; Osaka, A. Self-ordering of cell configuration of anodic porous alumina with large-size pores in phosphoric acid solution. *Jpn. J. Appl. Phys,* **1998**, *37*, 1340-1342.
[http://dx.doi.org/10.1143/JJAP.37.L1340]

[19] Ebihara, K.; Takahashi, H.; Nagayama, M. Structure and density of anodic oxide films formed on aluminium in oxalic acid solutions. *J. Metal Surf. Fin. Soc. Jpn.,* **1983**, *34*, 548-553.

[20] Csok'an, P.; Sc, C.C. Hard anodizing: Studies of the relation between anodizing conditions and the growth and properties of hard anodic oxide coatings. *Electroplat. Metal Finish.,* **1962**, *15*, 75-82.

[21] Lee, W.; Ji, R.; Gösele, U.; Nielsch, K. Fast fabrication of long-range ordered porous alumina membranes by hard anodization. *Nat. Mater.,* **2006**, *5*(9), 741-747.
[http://dx.doi.org/10.1038/nmat1717] [PMID: 16921361]

[22] Biring, S.; Tsai, K.T.; Sur, U.K.; Wang, Y.L. Electrochemically replicated smooth aluminum foils for anodic alumina nanochannel arrays. *Nanotechnology,* **2008**, *19*(1), 015304-015308.
[http://dx.doi.org/10.1088/0957-4484/19/01/015304] [PMID: 21730530]

[23] Asoh, H.; Nishio, K.; Nakao, M.; Tamamura, T.; Masuda, H. Conditions for Fabrication of Ideally Ordered Anodic Porous Alumina Using Pre-textured Al. *J. Electrochem. Soc,* **2001**, *148*, B152-B156.

[24] Liu, C.Y.; Datta, A.; Wang, Y.L. Ordered anodic alumina nanochannels on focused-ion-beam-prepatterned aluminum surface. *Appl. Phys. Lett.,* **2001**, *78*, 120-122.
[http://dx.doi.org/10.1063/1.1335543]

[25] Biring, S.; Tsai, K.T.; Sur, U.K.; Wang, Y.L. High speed fabrication of aluminum nanostructures with 10 nm spatial resolution by electrochemical replication. *Nanotechnology,* **2008**, *19*(35), 355302-355305.
[http://dx.doi.org/10.1088/0957-4484/19/35/355302] [PMID: 21828842]

[26] Yasui, K.; Nishio, K.; Nunokawa, H.; Masuda, H. Ideally ordered anodic porous alumina with sub-50 nm hole intervals based on imprinting using metal molds. *J. Vac. Sci. Technol. B,* **2005**, *23*, L9-L11.
[http://dx.doi.org/10.1116/1.1941247]

[27] Asoh, H.; Nishio, K.; Nakao, M.; Yokoo, A.; Tamamura, T.; Masuda, H. Fabrication of ideally ordered anodic porous alumina with 63 nm hole periodicity using sulfuric acid. *J. Vac. Sci. Technol. B,* **2001**, *19*, 569-572.
[http://dx.doi.org/10.1116/1.1347039]

[28] Lee, W.; Park, S.J. Porous anodic aluminum oxide: anodization and templated synthesis of functional

nanostructures. *Chem. Rev.,* **2014**, *114*(15), 7487-7556.
[http://dx.doi.org/10.1021/cr500002z] [PMID: 24926524]

[29] Schwarzacher, W.; Kasyutich, W. O. I.; Evans, P. R.; Darbyshire, M. G.; Yi, G. Fedosyuk, V.M.; Rousseaux, F.; Cambril, E.; Decanini, D. Metal nanostructures prepared by template electrodeposion. *J. Magn. Magn. Mater.,* **1999**, *199*, 185-190.

[30] van der Zande, B.M.I.; Bo¨hmer, M.R.; Fokkink, L.G.J.; Scho¨nenberger, C. Colloidal dispersions of gold rods: synthesis and optical properties. *Langmuir,* **2000**, *16*, 451-458.
[http://dx.doi.org/10.1021/la9900425]

[31] Zelenski, C.M.; Hornyak, G.L.; Dorhout, P.K. Synthesis and characterization of CdS particles within a nanoporous aluminum oxide template. *Nanostruct. Mater.,* **1997**, *9*, 173-176.
[http://dx.doi.org/10.1016/S0965-9773(97)00046-9]

[32] Pen˜a, D.J.; Mbindyo, J.K.N.; Carado, A.J.; Mallouk, T.E.; Keating, C.D.; Razavi, B.; Mayer, T.S. Template Growth of Photoconductive Metal-CdSe-Metal Nanowires. *J. Phys. Chem. B,* **2002**, *106*, 7458-7462.
[http://dx.doi.org/10.1021/jp0256591]

[33] Li, A.P.; Mu¨ller, F.; Birner, A.; Nielsch, K.; Go¨sele, U. Adv. Mater.~Weinheim. *Ger.,* **1999**, *11*, 483.

[34] Sberveglieri, G.; Murri, R.; Pinto, N. Characterization of porous Al_2O_3-SiO_2/Si sensor for low and medium humidity ranges. *Sens. Actuators B Chem.,* **1995**, *23*, 177-180.
[http://dx.doi.org/10.1016/0925-4005(94)01270-R]

[35] Kukhta, A.V.; Gorokh, G.G.; Kolesnik, E.E.; Mitkovets, A.I.; Taoubi, M.I.; Koshin, Y.A.; Mozalev, A.M. Nanostructured alumina as cathode of organic light emitting devices. *Surf. Sci.,* **2002**, *593*, 507-510.
[http://dx.doi.org/10.1016/S0039-6028(02)01320-1]

[36] Masuda, H.; Ohya, M.; Asoh, H.; Nakao, M.; Nohtomi, M.; Tamamura, T. Photonic crystal using anodic porous alumina. *Jpn. J. Appl. Phys., ,* **1999**, *38*, 1403-1405.
[http://dx.doi.org/10.1143/JJAP.38.L1403]

[37] Santos, A. Nanoporous anodic alumina photonic crystals: Fundamentals, developments and perspectives. *J. Mater. Chem. C Mater. Opt. Electron. Devices,* **2017**, *5*, 5581-5599.
[http://dx.doi.org/10.1039/C6TC05555A]

[38] Law, C.S.; Lim, S.Y.; Abell, A.D.; Voelcker, N.H.; Santos, A. Nanoporous Anodic Alumina Photonic Crystals for Optical Chemo- and Biosensing: Fundamentals, Advances, and Perspectives. *Nanomaterials (Basel),* **2018**, *8*(10), 1-8.
[http://dx.doi.org/10.3390/nano8100788] [PMID: 30287772]

[39] Santos, A.; Pereira, T.; Law, C.S.; Losic, D. Rational engineering of nanoporous anodic alumina optical bandpass filters. *Nanoscale,* **2016**, *8*(31), 14846-14857.
[http://dx.doi.org/10.1039/C6NR03490J] [PMID: 27453573]

[40] Manohar, C.; Andrea, T.; Anisha, G.; Gobind, D.; Zaccaria, R.P.; Roman, K.; Eliana, R. Chirumamilla, M.; Toma, A.; Gopalakrishnan, A.; Das, G.; Zaccaria, R.P.; Krahne, R.; Rondanina, E.; Leoncini, M.; Liberale, C.; De Angelis, F.; Di Fabrizio, E. 3D nanostar dimers with a sub-10-nm gap for single-/few-molecule surface-enhanced raman scattering. *Adv. Mater.,* **2014**, *26*(15), 2353-2358.
[http://dx.doi.org/10.1002/adma.201304553] [PMID: 24452910]

[41] Gracie, K.; Correa, E.; Mabbott, S.; Dougan, J.A.; Graham, D.; Goodacre, R.; Faulds, K. Simultaneous detection and quantification of three bacterial meningitis pathogens by SERS. *Chem. Sci. (Camb.),* **2014**, *5*, 1030-1040.
[http://dx.doi.org/10.1039/C3SC52875H]

[42] Domke, K.F. Surface Enhanced Raman Spectroscopy. Analytical, Biophysical and Life Science Applications. Edited by Sebastian Schlücker. Angew. In: *Chem. Int. Ed*; , **2011**; 50, p. 8226.

[43] Harper, M.M.; Dougan, J.A.; Shand, N.C.; Graham, D.; Faulds, K. Detection of SERS active labelled

DNA based on surface affinity to silver nanoparticles. *Analyst (Lond.),* **2012**, *137*(9), 2063-2068.
[http://dx.doi.org/10.1039/c2an35112a] [PMID: 22434199]

[44]   Kneipp, K.; Dasari, R.R.; Wang, Y. Near-Infrared Surface-Enhanced Raman-Scattering (Nir Sers) on Colloidal Silver and Gold. *Appl. Spectrosc.,* **1994**, *48*, 951-955.
[http://dx.doi.org/10.1366/0003702944029776]

[45]   Michota, A.; Bukowska, J. Surface-enhanced Raman scattering (SERS) of 4-mercaptobenzoic acid on silver and gold substrates. *J. Raman Spectrosc.,* **2003**, *34*, 21-25.
[http://dx.doi.org/10.1002/jrs.928]

[46]   Alvarez-Puebla, R.A.; Dos Santos Júnior, D.S.; Aroca, R.F. Surface-enhanced Raman scattering for ultrasensitive chemical analysis of 1 and 2-naphthalenethiols. *Analyst (Lond.),* **2004**, *129*(12), 1251-1256.
[http://dx.doi.org/10.1039/b410488a] [PMID: 15565227]

[47]   Fabris, L.; Dante, M.; Nguyen, T-Q.; Tok, J.B.H.; Bazan, G.C. SERS Aptatags: New Responsive Metallic Nanostructures for Heterogeneous Protein Detection by Surface Enhanced Raman Spectroscopy. *Adv. Funct. Mater.,* **2008**, *18*, 2518-2525.
[http://dx.doi.org/10.1002/adfm.200800301]

[48]   Millstone, J.E.; Park, S.; Shuford, K.L.; Qin, L.; Schatz, G.C.; Mirkin, C.A. Observation of a quadrupole plasmon mode for a colloidal solution of gold nanoprisms. *J. Am. Chem. Soc.,* **2005**, *127*(15), 5312-5313.
[http://dx.doi.org/10.1021/ja043245a] [PMID: 15826156]

[49]   Indrasekara, A.S.D.S.; Meyers, S.; Shubeita, S.; Feldman, L.C.; Gustafsson, T.; Fabris, L. Gold nanostar substrates for SERS-based chemical sensing in the femtomolar regime. *Nanoscale,* **2014**, *6*(15), 8891-8899.
[http://dx.doi.org/10.1039/C4NR02513J] [PMID: 24961293]

[50]   Hao, F.; Nehl, C.L.; Hafner, J.H.; Nordlander, P. Plasmon resonances of a gold nanostar. *Nano Lett.,* **2007**, *7*(3), 729-732.
[http://dx.doi.org/10.1021/nl062969c] [PMID: 17279802]

[51]   Talley, C.E.; Jackson, J.B.; Oubre, C.; Grady, N.K.; Hollars, C.W.; Lane, S.M.; Huser, T.R.; Nordlander, P.; Halas, N.J. Surface-enhanced Raman scattering from individual au nanoparticles and nanoparticle dimer substrates. *Nano Lett.,* **2005**, *5*(8), 1569-1574.
[http://dx.doi.org/10.1021/nl050928v] [PMID: 16089490]

[52]   Schütz, M.; Schlücker, S. Molecularly linked 3D plasmonic nanoparticle core/satellite assemblies: SERS nanotags with single-particle Raman sensitivity. *Phys. Chem. Chem. Phys.,* **2015**, *17*(37), 24356-24360.
[http://dx.doi.org/10.1039/C5CP03189C] [PMID: 26329892]

[53]   Romo-Herrera, J.M.; Alvarez-Puebla, R.A.; Liz-Marzán, L.M. Controlled assembly of plasmonic colloidal nanoparticle clusters. *Nanoscale,* **2011**, *3*(4), 1304-1315.
[http://dx.doi.org/10.1039/c0nr00804d] [PMID: 21229160]

[54]   Chen, G.; Wang, Y.; Yang, M.; Xu, J.; Goh, S.J.; Pan, M.; Chen, H. Measuring ensemble-averaged surface-enhanced Raman scattering in the hotspots of colloidal nanoparticle dimers and trimers. *J. Am. Chem. Soc.,* **2010**, *132*(11), 3644-3645.
[http://dx.doi.org/10.1021/ja9090885] [PMID: 20196540]

[55]   Wang, H. H.; Liu, C. Y.; Wu, S. B.; Liu, N. W.; Peng, C. Y.; Chan, T. H.; Hsu, C. F.; Wang, J.K.; Wang, Y.L. Highly raman-enhancing substrates based on silver nanoparticle arrays with tunable sub-10 nm gaps. *Adv. Mater.,* **2006**, *18*, 491-495.
[http://dx.doi.org/10.1002/adma.200501875]

[56]   Liu, T.T. A high speed detection platform monitoring antibiotic-induced chemical changes in bacteria cell wall. *PLoS One,* **2009**, *4*, 1-9.
[http://dx.doi.org/10.1371/journal.pone.0005470] [PMID: 19421405]

[57]   Malinovskis, U.; Poplausks, R.; Erts, D.; Ramser, K.; Tamulevicius, S. Malinovskis, U.; Poplausks, R.; Erts, D.; Ramser, K.; Tamulevičius, S.; Tamulevičienė, A.; Gu, Y.; Prikulis, J. High-density plasmonic nanoparticle arrays deposited on nanoporous anodic alumina templates for optical sensor applications. *Nanomaterials (Basel),* **2019**, *9*(4), 531.
[http://dx.doi.org/10.3390/nano9040531] [PMID: 30987127]

[58]   Hao, Q.; Huang, H.; Fan, X.; Yin, Y.; Wang, J.; Li, W.; Qiu, T.; Ma, L.; Chu, P.K.; Hao, Q.; Huang, H.; Fan, X.; Yin, Y.; Wang, J.; Li, W.; Qiu, T.; Ma, L.; Chu, P.K.; Schmidt, O.G. Controlled patterning of plasmonic dimers by using an ultrathin nanoporous alumina membrane as a shadow mask. *ACS Appl. Mater. Interfaces,* **2017**, *9*(41), 36199-36205.
[http://dx.doi.org/10.1021/acsami.7b11428] [PMID: 28948758]

[59]   Tabrizi, M.A.; Ferré-Borrull, J.; Marsal, L.F. Highly sensitive remote biosensor for the determination of lead (II) ions by using nanoporous anodic alumina modified with DNAzyme. *Sens. Actuators B Chem.,* **2020**, *321*, 128314.
[http://dx.doi.org/10.1016/j.snb.2020.128314]

[60]   Amouzadeh Tabrizi, M.; Ferré-Borrull, J.; Marsal, L.F. Remote biosensor for the determination of trypsin by using nanoporous anodic alumina as a three-dimensional nanostructured material. *Sci. Rep.,* **2020**, *10*(1), 2356.
[http://dx.doi.org/10.1038/s41598-020-59287-7] [PMID: 32047212]

[61]   Tabrizi, M.A.; Ferré-Borrull, J.; Marsal, L.F. Remote sensing of Salmonella-specific DNA fragment by using nanoporous alumina modified with the single-strand DNA probe. *Sens. Actuators B Chem.,* **2020**, *304*, 127302.
[http://dx.doi.org/10.1016/j.snb.2019.127302]

[62]   Belwalkar, A.; Grasing, E.; Van Geertruyden, W.; Huang, Z.; Misiolek, W.Z. Effect of processing parameters on pore structure and thickness of anodic aluminum oxide (AAO) tubular membranes. *J. Membr. Sci.,* **2008**, *319*(1-2), 192-198.
[http://dx.doi.org/10.1016/j.memsci.2008.03.044] [PMID: 19578471]

[63]   Shi, X.; Xiao, A.; Zhang, C.; Wang, Y. Growing covalent organic frameworks on porous substrates for molecule-sieving membranes with pores tunable from ultra- to nanofiltration. *J. Membr. Sci.,* **2019**, *576*, 116-122.
[http://dx.doi.org/10.1016/j.memsci.2019.01.034]

[64]   Najma, B.; Kasi, A.K.; Kasi, J.K.; Akbar, A.; Bokhari, S.M.A.; Stroe, I.R. ZnO/AAO photocatalytic membranes for efficient water disinfection: Synthesis, characterization and antibacterial assay. *Appl. Surf. Sci.,* **2018**, *448*, 104-114.
[http://dx.doi.org/10.1016/j.apsusc.2018.04.063]

[65]   Aghili, H.; Hashemi, B.; Bahrololoom, M.E.; Jahromi, S.A.J. Fabrication and characterization of nanoporous anodic alumina membrane using commercial pure aluminium to remove Coliform bacteria from wastewater. *Process. Appl. Ceram.,* **2019**, *13*, 235-243.
[http://dx.doi.org/10.2298/PAC1903235A]

[66]   Amouzadeh Tabrizi, M.; Ferre-Borrull, J.; Marsal, L.F. Advances in optical biosensors and sensors using nanoporous anodic alumina. *Sensors (Basel),* **2020**, *20*(18), 5068.
[http://dx.doi.org/10.3390/s20185068] [PMID: 32906635]

[67]   Chen, Q.; Wang, Y.; Zheng, M.; Fang, H.; Meng, X. Nanostructures confined self-assembled in biomimetic nanochannels for enhancing the sensitivity of biological molecules response. *J. Mater. Sci. Mater. Electron.,* **2018**, *29*, 19757-19767.
[http://dx.doi.org/10.1007/s10854-018-0101-2]

[68]   Chen, Z.; Ni, S.; Han, S.; Crawford, R.; Lu, S.; Wei, F.; Chang, J.; Wu, C.; Xiao, Y. Nanoporous microstructures mediate osteogenesis by modulating the osteo-immune response of macrophages. *Nanoscale,* **2017**, *9*(2), 706-718.
[http://dx.doi.org/10.1039/C6NR06421C] [PMID: 27959374]

[69]   Rahman, S.; Atkins, G.J.; Findlay, D.M.; Losic, D. Nanoengineered drug releasing aluminium wire implants: a model study for localized bone therapy. *J. Mater. Chem. B Mater. Biol. Med.,* **2015**, *3*(16), 3288-3296.
[http://dx.doi.org/10.1039/C5TB00150A] [PMID: 32262323]

[70]   Chen, X. Z.; Shamsudhin, N.; Huang, T.; Ozkale, B.; Li, Q.; Siringil, E.; Mushtaq, F. Chen, X.Z.; Hoop, M.; Shamsudhin, N.; Huang, T.; Özkale, B.; Li, Q.; Siringil, E.; Mushtaq, F.; Di Tizio, L.; Nelson, B.J.; Pané, S. Hybrid magnetoelectric nanowires for nanorobotic applications: fabrication.magnetoelectric coupling, and magnetically assisted *in vitro* targeted drug delivery. *Adv. Mater.,* **2017**, *29*(8), 1605458.
[http://dx.doi.org/10.1002/adma.201605458] [PMID: 27943524]

[71]   Li, J. EstebanFernandezde_Avila, B.; Gao, W.; Zhang, L.; Wang, J. Micro/nanorobots for biomedicine: delivery, surgery, sensing,and detoxification. *Sci. Robot.,* **2017**, *2*, 1-10.
[http://dx.doi.org/10.1126/scirobotics.aam6431]

[72]   Wang, B.; Kostarelos, K.; Nelson, B.J.; Zhang, L. Trends in micro-/nanorobotics:materials development, actuation, localization, and system integration for biomedical applications. *Adv. Mater.,* **2021**, *33*(4), e2002047.
[http://dx.doi.org/10.1002/adma.202002047] [PMID: 33617105]

[73]   Vyskocil, J. MayorgaMartinez, C. C.; Jablonska, E.; Novotny, F.; Ruml, T.; Pumera, M. Cancer cells microsurgery *via* asymmetric bent surface Au/Ag/Ni microrobotic scal-pels through a transversal rotating magnetic field. *ACS Nano,* **2020**, *14*, 8247-8256.
[http://dx.doi.org/10.1021/acsnano.0c01705] [PMID: 32544324]

[74]   Oshima, H.; Kikuchi, H.; Nakao, H.; Itoh, K. Detecting dynamic signals of ideally ordered nanohole patterned disk media fabricated using nanoimprint lithography. *Appl. Phys. Lett.,* **2007**, *91*, 022508-022510.
[http://dx.doi.org/10.1063/1.2757118]

[75]   Rajeev, G.; Prieto Simon, B.; Marsal, L.F.; Voelcker, N.H. Advances in Nanoporous Anodic Alumina-Based Biosensors to Detect Biomarkers of Clinical Significance: A Review. *Adv. Healthc. Mater.,* **2018**, *7*(5), 1-18.
[http://dx.doi.org/10.1002/adhm.201700904] [PMID: 29205934]

CHAPTER 5

PIXE/PIGE Measurements of Archaeological Glass, its Conceptualization and Interpretation: A Case Study

Roman Balvanović[1,*] and **Žiga Šmit[2]**

[1] *Vinča Institute of Nuclear Sciences, National Institute of Serbia, University of Belgrade, Belgrade, Serbia*

[2] *Faculty of Mathematics and Physics, Jožef Stefan Institute, University of Ljubljana, Ljubljana, Slovenia*

Abstract: Reliable scientific answers to questions posed by social sciences, like archeology, to exact sciences, like physics or chemistry, depend not only on meaningfully posed questions, well-selected and pretreated samples and accurate and precise measurements, but also on an area of interpretation that exists between the two fields. This interdisciplinary area consists of many representations of measurement data, notions, and concepts that evolve through solving particular problems. However, this set of concepts is not always determinate, clear and consistent, obscuring the problem and obstructing the interpretation of results. The chapter explains this starting with a concrete example of measurements of archaeological glass using simultaneous PIXE/PIGE measurements, explaining general and technical details of measurements, and proceeds to show how the measurements are treated, processed, and displayed in ways to comply with the concepts, interpret the results, and provide adequate answers to the questions posed by archeology. The chapter also offers some possible improvements through a few novel concepts and ways of interpretation.

Keywords: Compositional group, Conceptualization, Geological class, Kernel density estimate, PIGE, PIXE, Principal component analysis, Rare earth patterns, Roman glass.

INTRODUCTION

The chapter focuses on the most important issues and approaches regarding the measurement and interpretation of analytical data in Archaeometry, illustrated by PIXE/PIGE measurements of ancient glass. It describes the most important con-

* **Corresponding author Roman Balvanović:** Vinča Institute of Nuclear Sciences, National Institute of Serbia, University of Belgrade, Belgrade, Serbia; E-mail: broman@vinca.rs

Sibel A. Ozkan (Ed.)

cepts encountered in efforts how to understand and interpret the measurements that are meaningful for both the analyst and the archaeologist.

The chapter introduces several novelties. The equations for how to analyze glass with PIXE-PIGE without measuring the proton number have not been published before. There is also further improvement in the correction for geometrical effects. New ways to use principal component analysis (PCA) in data interpretation are suggested, enabling the depiction and comparison of large amounts of data in multivariate space. A step toward the identification of entities and relations in the interdisciplinary space between archaeological concepts and scientific measurements is made with the desire to lay a foundation for a formal interdisciplinary theory. All of this will be illustrated in the case study of the collection of eighty pieces of glass fragments (window panes, glass vessels, lamps, glass adornments) from the sixth century AD Byzantine settlement of Jelica, Serbia.

SIMULTANEOUS PIXE AND PIGE MEASUREMENTS OF ARCHAEOLOGICAL GLASS

How to analyze the chemical composition of glass? Glass is composed of light and heavy elements that, in principle, require different techniques for their determination. However, the metals in glass are in an oxide form with a known stoichiometric ratio, so it is then sufficient to determine the glass metal content. Though specific glasses may contain boron and fluorine, it is normally assumed that the analysis shall extend from (including) sodium onwards. The elements of $Z>10$ emit measurable X-rays, so X-ray fluorescence-based techniques may provide several advantages: they cover a large range of elements and are nondestructive, which is important when dealing with the objects of cultural heritage. For excitation of X-rays, irradiation with photons (X-ray fluorescence, XRF), electrons in the electron microscope (energy-dispersive spectroscopy, EDS) or charged particles (proton-induced X-ray emission, PIXE) can be used. The X-rays of sodium have an energy of 1.04 keV, so they absorb already in a few micrometers of material, including the irradiated sample itself. Corrosion processes that leach sodium out of the surface layer, causing sodium depletion, further afflict these shallow regions. The measurements are then preferably made in vacuum (EDS, PIXE) using thin-window X-ray detectors with no other absorbing material. For measurements at ambient pressure, helium flux may be used (XRF, in-air PIXE). For controlling the X-ray self-absorption, the glass object is usually sampled, and the sample is mirror-polished with abrasives and diamond paste. This is especially required for the measurements in an electron microscope, as the range of electrons of a few 10 keV energy is typically a few micrometers. Protons in the MeV energy range are much more penetrative and can reach several 10 μm deep into the object; however, self-absorption of X-rays defines the mean X-ray production range of about 10 μm for medium Z elements. It is sufficient that an unprepared glass surface can be used in PIXE measurement if the iridescent layer is peeled off. Another advantage of the proton beam is the small bremsstrahlung background. Bremsstrahlung X-rays are produced by accelerated charges: these are primary and displaced electrons in the case of the electron beam, and the resulting radiation can completely screen the characteristic X-rays of the elements heavier than iron. A complete analysis of a glass sample can be done by combining EDS with XRF. In vacuum, PIXE measurement is done in the same way as EDS in the electron microscope; the sample needs to be polished because of the self-absorption of sodium X-rays, though the sensitivity levels are significantly lower due to the smaller background, so heavier elements can be detected as well.

PIXE can also be applied in a way that the proton beam is extracted into the air through a thin window made of thin metal of plastic foils; the current material is silicon nitride (Si_3N_4) of sub-micrometer thickness. The objects are then analyzed in situ without limitation of their size. The measurements are almost non-destructive, as the radiation damage is minimal, and preparation of the measuring spots can be avoided, as some clean plane spots are easily found on the whole object. Several facilities apply helium flush in the irradiated area, which reduces proton energy loss between the exit window and target and, at the same time, reduces the absorption of soft X-rays [1]. For the apparatus in Ljubljana, we decided on ambient aerial atmosphere with the purpose to save helium, as helium supplies in the world are diminishing. This implied limitation for soft X-rays, which are extensively attenuated in air, shifting the lightest element to be detected to silicon. Important glass elements of sodium, magnesium and aluminum thus became invisible. Their detection is nevertheless possible by another spectroscopic method, detection of gamma rays induced by inelastic proton scattering (proton-induced gamma emission, PIGE). Gamma rays are more penetrative than X-rays of low-Z elements, so sodium is analyzed below the depths afflicted by corrosion processes. Detection of sodium by PIGE is then practiced also in systems using vacuum or He atmosphere [2]. For measurements in air, PIGE can also be used for the analysis of magnesium and aluminum. The side effect of measurement in the air is the detection of X-rays from atmospheric argon, which can be exploited as well as an indicator of the impact of proton number or for monitoring the influence of experimental geometry on self-attenuation of X-rays.

Principles of PIXE

The basic comprehension of X-ray analytical methods is simple: irradiation of analyte atoms removes electrons from their inner shells, while the resulting vacancy recombination is, to a certain probability, accompanied by the emission of characteristic X-rays. The X-ray energy is an indicator of the unknown element, while the number of X-rays is proportional to the elemental concentration. However, such a simplified approach would only be valid if the target atoms were isolated and independent of each other. In a real experiment, the glass analysis is always performed on a thick target whose composition strongly influences the number of induced X-rays. Besides self-absorption of X-rays, the protons gradually loose energy along their path into the sample, which in turn reduces the ionization cross section that quantitatively determines the vacancy production. For N_p protons hitting a thick target with energy E_1, the number of X-rays emitted by a specific element i and detected by an X-ray detector with a solid angle $\Delta\Omega$ is given by:

$$Y_i = \frac{\Delta\Omega}{4\pi} N_p N_A \frac{\varepsilon_i \eta_i}{M_i} x_i \int_0^{E_1} \frac{\sigma_i^x(E)}{S(E)} \exp(-\mu_i \xi(E)) dE \qquad (1)$$

The other parameters in (1) are as follows: The X-ray detector has a counting efficiency ε_i, while the transmission of absorbers between the target and detector (including airgap) is η_i. The element i is present at a mass fraction x_i and has atomic mass M_i. N_A is Avogadro's number. The integral in (1) is usually called the thick target factor and contains X-ray production cross section σ_i^x, proton stopping power S (in units of MeV cm^2/g) and X-ray attenuation coefficient μ_i. The quantity ξ is the length the induced X-rays travel in the target. For a plane-smooth target with a well-defined proton impact at an angle α (with respect to the surface normal) and X-ray take-off angle ψ, it is given by a simple geometrical expression:

$$\xi = z \frac{\cos \alpha}{\cos \psi}, \qquad z = \int_E^{E_1} \frac{dE}{\rho \, S(E)} \qquad (2)$$

Where, z is the distance along which the proton energy reduces from E_1 to E. The approximation (2) is more critical than one can imagine, as the angles α and ψ can easily deviate from their nominal values. The target is usually placed into the beam by subjective feeling, which may result in an uncontrollable rotation of the sample. The other effect that may violate the validity of eq. (2) is surface roughness, which shifts the distribution of the lengths ξ to higher values. As a simple approximation to partly compensate this effect, one can calculate eqs. (1, 2) with a somewhat larger value of the angle ψ [3].

According to eq. (1), it is also necessary to measure the number of incoming protons N_p. In our case, we measure the proton current by a chopper that intersects the proton beam in a vacuum. The chopper arms bear a thin gold foil, and the hitting protons elastically recoil according to Rutherford's law. As the mass of gold atoms is much higher than the masses of the underlying substrate, the gold scattering peak is well separated from the others, so its area is easily determined. Alternative methods apply detection of the signal from the exit window. We also developed the method that exploits the X-ray signal of argon atoms from the air gap between the exit window and target. The advantage of the method is that argon X-rays are always present (except the detector is equipped with a very thick absorber), and the ratio between target and argon X-ray yields is not sensitive to detector dead time effects. The disadvantage is, however, that the Ar yield needs to be modelled and corrected for secondary fluoresce, induced by target X-rays with energies just above the Ar absorption edge (potassium, calcium). Experimentally, the airgap distance needs to be kept constant, which we attain by mechanical spacers. It is also necessary to estimate the fraction of the Ar yield that is geometrically shadowed by the exit nozzle [4]. Measurements of proton number can be avoided in specific cases when the sum of mass fractions can be normalized to a certain number, like 100% in metal alloys with Z>10. In glass, it is possible to set to 100% the sum of metal oxides. The normalization constant is actually the constant factor in eq. (1),

$$A = \frac{\Delta\Omega}{4\pi} N_p N_A \tag{3}$$

Principles of PIGE

For in-air PIXE, the X-rays of sodium, magnesium and aluminum are drastically attenuated in the air gap between the target and detector. Using a few cm long air gap is recommended to stop the scattered protons to reach and deteriorate the detector crystal. For normalization, according to Ar, about 5 cm air gap is ideal as the argon yield is compensated for air-pressure variations: increased pressure means more argon, but also stronger attenuation. We have therefore decided that the three elements are rather determined according to the characteristic gamma-rays induced by inelastic nuclear scattering by protons. For determining the concentrations of Na, Mg and Al, we exploit the strongest gamma lines induced by (p, p') reactions. In Na, we choose the line at 440 keV; there are also two strong lines at 1634 and 1636 keV, but they are broader, and their shape is less similar to Gaussian. In Mg, the strongest line is at 585 keV. Its deficiency is the close proximity of a gamma line at 583 keV emitted by Tl^{208} on natural background. Its contribution is diminished by the shielding of the gamma detector with lead and by a sufficient count rate of proton-induced gamma rays. In Al, there are two strong lines at 844 and 1014 keV. Of the three elements, the most critical is the measurement of Mg, which directs the measuring time, necessary to collect sufficient counting statistics.In specific glasses, boron can be determined as well, for example through a line at 429 keV.

The induced gamma yields are determined by an equation similar to (1). However, as the gamma rays with energies above 100 keV are much more penetrative than X-rays, the attenuation term in (1) can be omitted. Further, as the predominant contribution to the thick target integral comes from isolated resonances, it is possible to extract the stopping power out of the integral and replace it with some mean value. In the so-called surface approximation, this mean value is calculated for the proton impact energy E_I. The gamma yield for the element i is then given by:

$$Y_i = \frac{\Delta\Omega\prime}{4\pi} N_p N_A \frac{\varepsilon_i \eta_i}{M_i S(E_1)} \; x_i \int_0^{E_1} \sigma_i^\gamma(E) \, dE \qquad (4)$$

The gamma rays are measured with a detector having a solid angle $\Delta\Omega$' and efficiency ε. We can see that the integral of the gamma-production cross-section σ_i^γ no more depends on the properties of the target; these are now only retained in the stopping power S. Unfortunately, the gamma production cross-sections are not known with an accuracy, such as for X-ray production cross sections, so the unknown concentrations are rather determined according to the standards, eliminating thus also unknown detector solid angle $\Delta\Omega$' and efficiency ε. From the standard concentration x_i^s, the unknown concentration is as follows:

$$x_i = \frac{Y_i}{Y_i^s} \frac{N_p^s}{N_p} \frac{S}{S^s} x_i^s \qquad (5)$$

Here, the suffix s refers to the measured gamma yield in the standard, respective proton number and stopping power.

Exploiting the normalizing principle, it is also possible to combine gamma and X-ray data and calculate the unknown concentrations without measuring the proton number N_p. The necessary condition is to execute the measurement in a way that the X-ray and gamma ray yields are induced by the same number of protons. Since two different detectors are used for both types of radiation, a correction for dead times is necessary. Extracting from (4) all parameters that are related to element i, we define the gamma calibration parameters as follows:

$$G_i = \frac{\Delta\Omega\prime}{\Delta\Omega} \frac{\varepsilon_i \eta_i}{M_i} \int_0^{E_1} \sigma_i^\gamma(E) \, dE \qquad (6)$$

In this way, it is possible to write eq. (4) in a shorthand form:

$$Y_i = \frac{\Delta\Omega}{4\pi} N_p N_A \frac{1}{S(E_1)} G_i x_i = A\, G_i\, \frac{x_i}{S(E_1)} \tag{7}$$

The parameters G_i are determined from the measurement on the glass standard. The normalization constant A follows from the measured X-ray yields, setting the sum of respective x_i to the value in the standard. In the unknown sample, the concentrations of light elements are then determined from eq. (7), and the concentrations of the heavy elements from eq. (1). Both sets are, however, uncertain for the normalization constant A of the sample, which is then obtained by setting the sum of all metal oxides to 100%.

Geometrical Effects in the Analysis – The Argon Signal

For X-rays, the deduced concentrations depend strongly on X-ray attenuation in the target. In a simple geometrical model (2), the attenuation depends on the angles α and ψ, which are not so easily controllable if the surface of the sample is irregular. The uncertainty of both angles reflects in the normalization constant A, which may, in turn, influence the evaluation of Na, Mg and Al from gamma yields according to eq. (7). A way to control the geometrical parameters is to rely on the argon yield induced in the air gap between the exit window and target. Two procedures have been developed, which are as follows:

a. The proton number N_p is measured. Since the gamma rays are practically not attenuated, we assume that only the concentrations of X-ray-determined elements are subject to geometrical effects. The normalization constant A is calculated from the Ar yield and compared to the one from the normalization of X-ray-based concentrations. The geometrical parameters are then adjusted manually until both values coincide with the accepted tolerance. In our earlier works, this was 20% [3].

b. The proton number N_p is not measured. In this case, we assume that the measurement on the standard, which has a large flat surface, yields reliable A and G_i, while A for the sample is subject to uncertainties. For the sample measurement, we then use the deduced A and calculate the apparent concentration of argon in the air. Again the geometrical parameters are varied until the argon concentration in the air (1.29 mass %) is reproduced within 1%.

Experimental Equipment

Measurements were executed at the Tandetron accelerator in Ljubljana, using the in-air proton beam of 3 MeV nominal energy (Fig. **1**). After passing an exit window of 200 nm Si_3N_4 and about 9 mm air gap, the energy on the target was about 2.93 MeV. The beam was focused by magnetic optics, consisting of two quadrupoles to the size of about 0.3 mm (sharp focusing was avoided in order to irradiate a larger area). The induced X-rays were detected by a Si(Li) detector of 160 eV resolution at 5.89 keV positioned approximately 6 cm from the target. The detector was equipped with a pinhole filter composed of 0.05 mm thick aluminum foil with a relative opening of 9%, which allowed the detection of elements in the energy window 1- 30 keV. Precise values of geometrical parameters were determined by measurements of pure elemental or simple chemical compound targets between SiO_2 and Zr. The induced gamma rays were detected by an intrinsic germanium detector of 40% efficiency (crystal dimensions 5 x 5 cm) positioned about 10 cm from the target. Glass NIST 620 was selected as the required standard. The current measurements were first performed by a chopper, however later analysis showed that a fraction of the beam hit the diaphragm before the Si_3N_4 window. We then rather relied on the procedure (b), which does not require the measured proton number but controls the geometrical parameters according to the deduced argon concentration. For the control, we repeated the evaluation by the procedure (a), regarding the argon yield as a measure for N_p without any other physical meaning. The agreement between the two methods was about 1%. For control measurements, we further measured the glass standards NIST 620, NIST 621 and BAM S005B as unknown samples. An agreement of 5% was obtained between the nominal and measured concentrations of the main constituents and about 10% for trace elements.

Fig. (1). The external beamline at the Tandetron accelerator of the Jožef Stefan Institute (2021).

ANALYSIS AND REPRESANTATION OF DATA

For an analyst it is all in numbers: concentrations, percentages, ratios, standard deviations, correlations. For an archaeologist, all is about interpretations: where, when and how. How to connect these two so that they truly understand each other in a meaningful way? In particular, how to interpret analytical data in such a way that the interpretation fits into the notions of the archaeologist? And vice versa: how to define an archaeological question in such a way that it can be converted to an analytical task so that measurements bear information that is relevant to the posed question? An obvious and oversimplified answer would be to sit and talk the problem over. Nevertheless, however unavoidable this step might be, it is not sufficient. Simply, the two do not speak the same language. Thus, what is essential is an intermediary. We are talking about concepts, notions and strategies that are emerging gradually and painstakingly from efforts to overcome each particular issue. We will explain the most important of these concepts and methods in the interdisciplinary field of archeometry, illustrating them with examples from a concrete study of ancient glass samples by analytical measurements. In addition, we will introduce a few new concepts and make an effort to generalize the approach to the area of interpretation to a certain degree. To make a linguistic parallel, we will try to outline the main features of the "syntax and semantics" of interdisciplinary "language" that can be used for communications between analytical scientist and archaeologist.

When an archaeologist makes questions about an object of interest, *e.g.*, how this glass vessel is produced, where and when it is made, analyst stands before the immediate questions: what do I measure? What will the results mean? How do I present them to the archaeologist?

This is the point where he or she has to turn to notions that are neither strictly analytical nor archaeological, but interdisciplinary ones. However, that is not always a straightforward and a non-contradictory solution. For example, to answer the question about the technology of production, the analyst will have to stick to the notion of two different types of flux that were in use during the first millennium CE – natron flux and plant ash flux. By measuring elemental concentrations of sodium, potassium, and magnesium oxides of the glass bulk, he or she can tell which technology was used in its production, whether mineral flux or plant-ash flux was used.

To answer the question when the vessel is made, the analyst will have to use the concepts of a "compositional group" or a "production group" to which to compare his glass. By establishing "similarity" to a group from a known assemblage or, better, to a compositional group, the analyst can claim the rough dating of the glass that he or she is analyzing, by implying that the dating is the same as the dating of the known group that is compositionally similar.

Yet another important notion is provenance. This relates to another two concepts. The first is the concept of two-phased glass production, which is widely accepted in the literature concerning the first millennium CE glass production. This concept says that glass in this epoch was produced in two stages, raw-glass production in the area of abundant and high-quality resources, sand and natron, and the transportation of such raw glass across the wide region to secondary glass workshops for glass blowing. Using this concept, the question of provenance of a particular glass object converts to the question of provenance of raw glass. To answer this question, the analyst has to turn to yet other concepts that are related to geology. In essence, he has to determine geological markers that are characteristic of the area from which the raw material, sand or natron, supposedly comes. In the case of the Roman glass, these markers are isotopic ratios of telltale elements like strontium, hafnium or oxygen. Another marker is so-called rare-earths patterns, graphs of relative concentrations of these elements in Earth's crust that bear information about specific geological processes that are reflected in the sand elemental compositions.

However, the concepts analyst is using are sometimes vague because their meaning differs from case to case. This stems naturally from the fact that each study investigates a concrete and to an extent always-arbitrary set of objects. Another reason is that compositional groups often overlap to a non-negligible degree. Yet another reason is that a term can reflect several basic attributes of the collection - technology, local variations of sand due to different sand quarries, sampling of objects to analyze, samples cleaning, choice of measurement techniques *etc.* Therefore, there is a non-negligible amount of indeterminacy in using these concepts.

In addition, all the mentioned concepts are "static" in a sense that, by analyzing the composition, they try to relate it to the actual production furnace [5], or place, and thus view upon an assemblage and a location of production as belonging to more or less the same period of time. However, it is known that a particular sand quarry was not exploited for a very long time, typically a few years, before it was moved to another location. The reason for this is considered to be either exhaustion of easily available sand or exhaustion of fuel (wood) for firing the raw glass-furnaces. Location changes cause changes in the composition of sand, which changes the composition of raw glass, which reflects in the composition of glass objects. In this context, what does the notion of "compositional group" mean? Especially so if one takes into account a realistic assumption that the search for ever new sand quarries continued on and on.

These examples show that while these concepts and notions are important in interpreting the goals of an investigation, they are in some cases indeterminate, contradictory and insufficient. We will explore these concepts in more details through concrete examples, and offer some possible improvements through a few new concepts and interpretations. This is the main topic of the second part of this chapter.

Roman Glass, Technology and Types

Roman glass is composed of three main components, silica, soda and lime. Silica is a glass bulk material; soda serves as a metallurgical flux to reduce the melting temperature of silica. Lime serves as glass stabilizer. Two types of flux were in use during the first millennium CE, of mineral and plant origin [6 - 8]. Until the 8th century CE, these three components were added to batches within two components, sand and flux [9]. This "natron glass" is characterized by low levels of magnesia and potash, bellow 1.5% each. From the 8th century CE on, plant ash started to be used as a flux. This type of glass has considerably raised concentrations of potash, calcium and magnesium [10]. The most typical glass of the Roman Empire was the "Roman glass" (1st – mid 4th century AD), with extraordinarily stable average concentrations of sodium 16.63%, alumina 2.59%, lime 7.48%, iron 0.62% and titanium 0.13% . During the 4th century, a noticeable change happened in the glass compositions [11]. Several new glass types appeared. A glass with elevated concentrations of iron, manganese and titanium, termed HIMT appeared [12, 13]. They are divided to HIMT 1 and HIMT 2 depending on the concentrations of these elements [14]. Its provenance is considered to be Sinai [15]. The glass from the 4th century workshop in Jalame [9], higher in alumina (2.7%) and lime (8.79%) and lower in sodium (15.8%) than the Roman glass, is termed Levantine I [16]. The glass from 7th – 8th century furnaces in Bet Eliezer, lower in sodium (12.3%) and lime (7.36%) and higher in alumina (3.26%) than Levantine I, is termed Levantine II [16]. There are two groups of "Egyptian glass" in the 1st millennium, Egypt I (7th - 8th century) and Egypt II (8th - 9th century [17]). Compared to Levantine, Egypt I is richer in alumina (4.46%) and iron (1.79%), Egypt II is lower in alumina (2.53%) and higher in lime (9.53%) [18]. Foy *et al.* [16, 14] give their own classification of Late Antiquity Mediterranean compositions, describing ten different natron glass groups [19, 20].

It is widely accepted that Roman glass production was organized in two phases, primary glass production and glass objects manufacturing [21 - 23, 19, 24, 25]. Primary glass was produced in relative proximity to high-quality raw materials in the Eastern Mediterranean (Levantine/Egyptian sand and natron from Wadi Natrun in Egypt), and exported in lumps of raw glass through a complex trade-network to local glassmakers across the Empire.

Sample Selection Criteria

Archaeologists often find themselves in a situation that they do not have a collection that is sufficiently rich in numbers, forms and well preserved (which limits type identification). In addition, they are often interested in unusual, rare, or well-preserved objects, or object prominent for some specific attribute like color and form. This means that the assemblage selected for analysis may not be too representative of the glass collection found at the archaeological site. In addition, this, by implication, impairs comparison between different assemblages. Since these limits are inherent in any archaeological collection, there is no clear way out of this situation. This impact on the representativeness of an assemblage selected for analysis should be kept as low as possible by an effort to select objects of different purposes and different attributes. At the same time, assemblage should be as representative as possible, keeping in mind the overall collection. For example, types should include objects of everyday use like beakers and goblets, bottles, lamps, windowpane fragments, ceremonial objects like balsamarii, decorative objects like bracelets, pearls and adornments. Colors selected should range wide, from colorless to naturally colored bluish, greenish, yellowish, green, olive-green. Of particular interest are deliberately colored objects, made of blue, black or purple glass.

Sample Preparation

Due to prolonged contact with soil and moisture, archaeological glass surface is often covered with a thin layer of oxides, which is called iridization. The elemental composition of this layer is different from that of unaffected bulk of glass and thus it should be cleansed before measurement. Glass samples are generally cleansed in ultrasonic bath in presence of absolute ethanol for 5 – 10 minutes, when ultrasonic vibrations detach this often-fragile corrosion layer from the glass. The next step is grinding with super fine sand paper (size 1000 or more), and rinsing well with absolute ethanol.

The spots cleanest from corrosion layers should be selected for analysis. This depends on the ion-beam size used to measure the glass. The smaller the beam size, the easier it is to find a clean spot. However, one should have in mind that smaller beam size implies a less representative (average) elemental composition of the bulk. Thus, the choice of the beam size should be selected optimally between these two conflicting influences.

Data Preprocessing (Heteroscedascity)

Prior to analyzing the raw data, some data preprocessing should be done. Since the glass is viewed as comprised of oxides, the first step is to calculate the percentages of all elements whose relative concentrations were determined by PIXE/PIGE as oxides, with a total sum of 100%. This enables later compositional comparison between various glass assemblages.

The second step is to inspect if the data are heteroscedastic (if the variances are equal or not across oxide values). This is important because heteroscedasticity can affect multivariate analysis, such as Primary Component Analysis, which is often used in evaluating overall composition of a glass assemblage. One of the methods of minimizing the influence of heteroscedasticity on PCA diagrams is power transformation of data:

$$\hat{x}_{ij} = \tilde{x}_{ij} - \bar{\tilde{x}}_\iota, \text{ where } \tilde{x}_{ij} = \sqrt{x_{ij}} \tag{8}$$

Major, Minor and Trace Elements and their Interpretative Diagrams

Major oxides in ancient glass analysis are SiO_2 and Na_2O. They come into glass from sand and natron, respectively, constituting the major part of the glass bulk - their concentrations for the natron glass range roughly in the range of 65-75% and 15-20%, respectively. Silica and sodium concentrations tell of the technological recipe used in making raw glass, the ratio of glass-making sand and natron used as a flux.

Fig. (**2**) depicts bi-plot of two major technological parameters, SiO_2 and Na_2O, for the glass assemblages called HIMT 2 from Roman Britain and the glass assemblage from Jalame in Roman Palestine. The assemblages are broadly contemporary, dated to the fourth century CE.

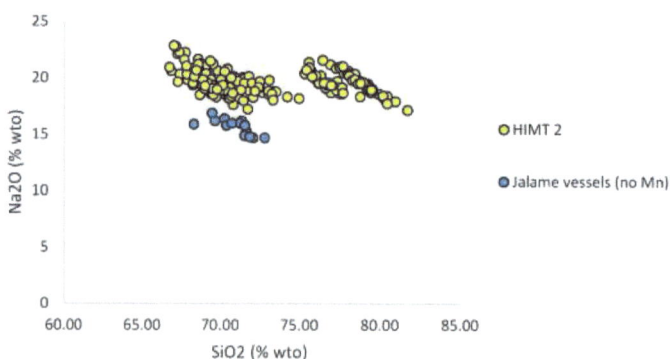

Fig. (2). Bi-plot showing basic technology of raw glass making: different mixtures of sand and natron for two assemblages, HIMT 2 and Jalame without deliberately added manganese. Note that for HIMT 2 SiO_2 measurements are not reported, but are calculated here so that sum of all oxides equals 100%.

Fig. (**2**) indicates that HIMT 2 assemblage seem to be produced with two different recipes. The group on the left is produced with silica to natron ratio roughly of 3.5:1 and the group on the right with proportion of roughly of 4:1. The interpretation might be that different producers used different recipes for production. In addition, the group with higher sand to natron ratio seems to be further composed of two slightly different subgroups, almost parallel to each other, and placed almost on straight lines. Interpretation of this might be that the two sets represent two separate batches, glass produced in two separate firings of the furnace.

It is interesting to note that the same bi-plot also shows some information on provenance of the glass, besides information on technological recipes. This is shown by the position of the assemblage from Jalame, consisting of vessels produced with no deliberately added manganese (as decolorizer). The provenance information is here implicit in the hypothesis that the amount of natron used in raw glass production is greater in Egypt than in Roman Palestine (another major producing area of the Roman glass, as testified by several archaeological finds of primary glass production centers in both of these regions, including furnaces). This is considered so because of the vast supply of natron from Wadi Natrun salt lakes near Kairo, and the presumed lower price of natron in Egypt than in Palestine. Thus, the Jalame glass from Palestine, with less access to natron from Egypt (or because of its supposed higher price in the Palestine) uses less natron for the same amount of silica sand (4.5:1). Fig. (**2**) demonstrates that a single bi-plot can provide several types of information – about provenance, technology and even some details about production process. This type of information is derived from underlying archaeological and technological concepts.

Minor elements of glass, in forms of oxides, are CaO, Al_2O_3, MgO, MnO, SO_3, P_2O_5, Cl, K_2O, TiO_2 and Fe_2O_3. Their origin is more complex. The most part of it comes from glass-producing sand (Al_2O_3, MgO, TiO_2, Fe_2O_3 and partially P_2O_5) and from lime (CaO). Some come from natron (SO_3, P_2O_5, Cl) and some are related to production process (SO_3, P_2O_5, Cl). Minor oxides that come from sand bear information about the geochemical processes that governed the formation of the rock from which the sand is derived. This information can be used to infer some general information about the provenance of the raw materials used in glass production.

The example of such inference is shown in Fig. (**3**) (up) which depicts concentrations of Al_2O_3 and CaO, derived from lighter minerals of sand, like feldspars, and lime. Levantine beach sands have higher concentrations of feldspars and lime from Egyptian sands. Their higher lime contents comes from calcium carbonate from seashells. The diagram distinguishes the origin of depicted glass groups to those of Levantine origin (Apollonia and Bet Eli'ezer) from those of Egyptian origin (Egypt I, Foy 3.2 and Foy 2.1). This diagram shows that glasses belong to distinguished groups of production of primary (raw) glass.

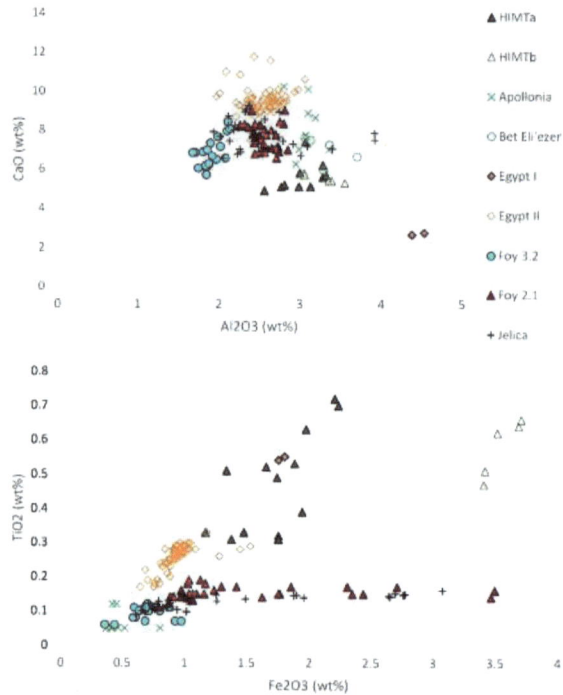

Fig. (3). Bi-plots showing concentrations of elements that are derived from light minerals of sand (up) and from heavy minerals of sand (down). Source [29].

Fig. (3) (down) depicts elements derived from heavy minerals of sand, iron and titanium. Egyptian glass-making sands are characterized by higher concentrations of heavy minerals, brought from Ethiopian highlands by river Nile and spread over Egyptian beaches by sea currents. Groups HIMTa, HIMTb and Egypt II have different correlations of iron and titanium, showing different heavy mineral suites of their respective sands.

The most important trace elements of glass are Co, Ni, Cu, Sr, Zr, Sn, Sb, Pb and Ba. They bear information on technological processes in glass production and about glass provenance. For example, recycling, a common production procedure where old crushed glass (cullet) was melted together with raw glass or other cullet, typically manifests itself in an increase of Ni, Cu, Zn, Sn and Pb. The cutoff values for concentrations of these elements that show recycling have to be calculated individually for every primary production group, *i.e.*, starting from the data for primary (raw) glass. This is because starting concentrations of these elements are different since they use different sands for production of raw glass. For example, values of Cu, Sb, Pb > 100 ppm are suggested by Foy *et al.* (2003) as cutoff values for her group 1.

Using meta-data of 246 glasses of type Foy3.2, we calculated the cutoff values of Foy 3.2 as 40 ppm for Co, 50 ppm for Ni and Zn, 60 ppm for Pb, and 70 ppm for Cu [26]. Above these levels, série 3.2-type of glass can be considered surely recycled, and below modestly recycled or unrecycled.

Main decolorizers are antimony and manganese. Antimony was in use in earlier epoch, until the end of the third century CE [27]. From the fourth century on, manganese started to replace antimony, presumably because of exhaustion of readily available antimony-bearing ores. Antimony is stronger decolorizer than manganese. For example, colorless glasses fully decolorized with antimony from Mala Kopašnica, Serbia [28] has Sb_2O_5:Fe_2O_3 ratio of 1.08, while colorless glass decolorized with manganese 3.55.

Antimony, prime decolorizer until the fourth century CE, can also serve as an indicator of recycling. From the fourth century CE on, manganese started to replace antimony as glass decolorizer. Since the practice of recycling continued in this epoch too, new colorless glass was often recycled with older, antimony-recycled cullet, leaving telltale traces of antimony in the new glass. We calculated that antimony cutoff for Foy 3.2 glass (otherwise manganese-decolorized) is 10 ppm (unpublished).

Other elements, otherwise in trace concentrations, are colorants, like cobalt, copper and lead. Cobalt is the strongest colorizer, giving dark blue color. For example, dark-blue decoration on beaker wall (sample 16) has only 580 ppm of cobalt. Copper is weaker colorizer than cobalt, and gives lighter blue color. For example, light blue fibula inlay from Jelica [29] (sample 24) has 1830 ppm of copper.

Glass with no added decolorizer, antimony of manganese, can be naturally colorless or colored, depending on the concentration of impurities in sand and redox conditions in furnace. True naturally colorless glasses were made of very clean sands, with low impurities. Concentrations of iron in such glass can be very small, like the glass 1b [30] that has 0.29% of iron and Fe_2O_3+TiO_2+MgO=0.81%. The samples on the Fig. (**4**) show glasses from Jelica in increasing concentrations of sand impurities. Almost colorless glass 26, made of sand with small concentrations of impurities, has 0.56% of Fe_2O_3, and Fe_2O_3+TiO_2+MgO = 1.11%. Glasses 36 and 10, produced with sands with high concentrations of impurities, have 3.65% and 3.91% of these oxides, respectively.

Fig. (4). Photographs of the glasses from Jelica (Serbia). Colors of glass range from colorless and almost colorless to naturally colored depending mostly on the amount of iron in glass-making sand and ratio of iron to deliberately added decolorizer (antimony and/or manganese). From left to right: colorless glass with no decolorizer (Fe_2O_3 = 0.33%), slightly colored glass, decolorized with manganese (Fe_2O_3 = 0.91%, MnO = 1.22%), naturally colored glass (Fe_2O_3 = 01.96%, MnO = 0.70%). Two deliberately colored glasses are colored with copper and cobalt. Source [29].

Minor elements can also be colorizers, like iron and manganese. For example, black bracelet bead (sample 32) from the sixth-century CE Jelica, Serbia, has 6.68% of iron. Manganese can give purple color under oxidizing conditions, like the cullet no. 649 from Jalame, with 2.79% of MnO.

Trace elements that are related to sand provenance are Rare Earths (REE). REEs found in sands, soils and sediments provide information that might give valuable insight into the geochemical process that controls rock and sediment formation [31]. Their correlations are indicative of specifics of local rock geochemistry, so they represent "fingerprints" of the sand used in glass production and are helpful in investigating the provenance of a glass or adherence of a glass to a particular production group. Fig. (5) shows correlations of REEs of two glass assemblages with different provenance, Egyptian [32] and Levantine [33]. Very high correlations of all Rare Earths, both light and heavy, are an indication of heavy sand minerals present in Egyptian sands (Fig. 5, up). Heavy minerals are brought by sea currents along the shores from the Nile delta towards the Levantine coast, so their concentrations decrease with distance from the Nile delta. The Levantine sands, from which Levantine I type of glass is produced, contain less heavy minerals and more light minerals like feldspars and lime, so their REE correlations are very different Fig. (5, down).

Specifics of the geochemistry of rocks from which sand is derived can also be investigated through REE patterns, which represent their concentrations normalized to the upper continental crust [34, 35]. Fig. (6) shows REE patterns for assemblages of Foy 2.1 type of glass from Visighotic Spain [36]. This type has a sub-variant consisting of glasses with higher concentrations of iron oxide, called Fe-rich Foy 2.1 or high-iron Foy 2.1 (Fig. 3). In the example in Fig. (6), the assemblage from Visighotic Spain is divided into three groups according to the iron oxide concentrations, and their respective REE patterns are depicted. To investigate the origin of iron in the glasses with higher concentrations, we group the glasses according to iron oxide concentrations in three groups, low iron (average of 0.94%), high iron (1.77%) and very high iron (2.63%). Fig. (**6**) shows that with the increase of iron from the low iron to the high iron group, concentrations of also REEs increase while keeping the same shape of REE pattern, *i.e.*, the differences in their respective heights related to iron concentrations (0.94% versus 1.77%). This shows the same origin of iron in both groups. The same is true with the increase from the high iron to the very high iron group, but with an exception of cerium and hafnium, which decrease with an increase in iron concentrations, indicating different type, and thus, likely different provenance of the mineral iron.

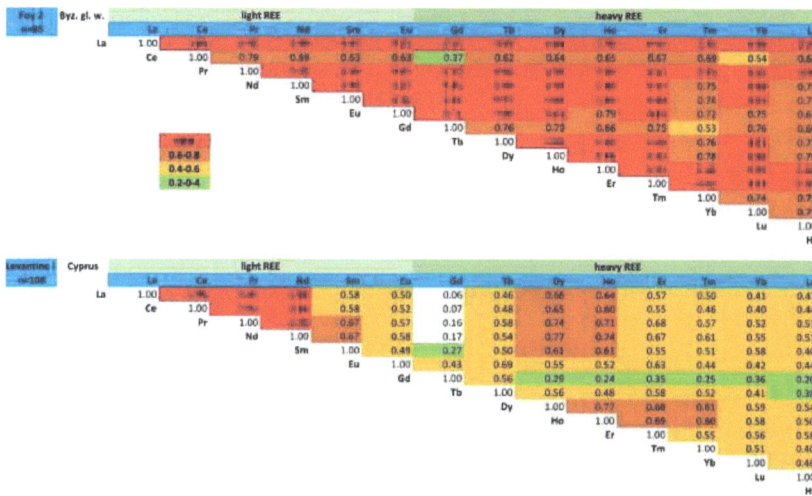

Fig. (5). Different correlations of Rare Earths in two glass assemblages show different geochemical characteristics of two glass-producing sands. Correlation of REEs in the collection of Byzantine glass weight from the British Museum and the Bibliothèque nationale de France, of type Foy 2 of Egyptian provenance (up) and collection from Cyprus, of type Levantine I of Levantine provenance (down).

Fig. (6). Trace element patterns of Fe-rich Foy 2.1 glasses from Visighotic Spain, grouped by iron concentrations. Values normalized to the concentrations of REEs in the upper continental crust [35].

Isotopic Ratios in Determining Glass Provenance

The transport of material across the Mediterranean is further reflected in the isotopic ratio of strontium and neodymium. Besides Nile, which contributes mostly to the sediments in the Mediterranean, Sahara winds and European rivers bring an important influx from Greece and Italy, as well as from Adriatic. The isotope ^{87}Sr is the product of beta decay of ^{87}Rb. Its abundance is measured with respect to ^{86}Sr, which is one of the natural strontium isotopes. The ratio $^{87}Sr/^{86}Sr$ varies from 0.703 in young basalts to 0.750 in granite. In the Mediterranean, it increases from 0.709 on the east to 0.718 on the west. The isotope ^{143}Nd is the product of alpha decay of ^{147}Sm. The characteristic quantity $\varepsilon_{Nd(0)}$ is given as a variation of the isotopic ratio of ^{143}Nd to natural isotope ^{144}Nd with respect to the ratio in the chondritic uniform reservoir. Across the Mediterranean, the values of $\varepsilon_{Nd(0)}$ decrease from -2 in the east to -12 in the west [37]. The reading of Pliny (XXXVI 194), who describes that the raw glass was also produced by the Volturnus River in Italy as well in Gallia and Hispania, inspired extensive research of siliceous sands in the Western Mediterranean [38 - 40]. For the origin of colorless glass, important results were obtained for the shipwrecks of Iulia Felix and Ouest Embiez [41]. The same isotopic ratio was found for the manganese-decolored glass of IF CL2 and Embiez 2, and for the antimony-decolored glass of IF CL1 and Embiez 1 (Fig. **7**), implying that both types of glass have different geographical origin; it is now considered that the antimony-decolored glass came from Egypt and manganese-decolored from the Levant. Other isotopic ratios are also used in provenance analysis, like hafnium and oxygen [42, 43]. Hafnium has shown a very good discriminator between Egyptian and Levantine sands.

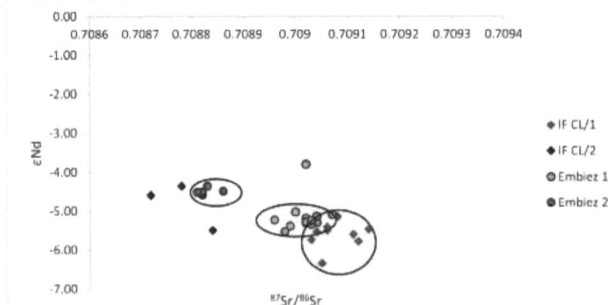

Fig. (7). Strontium and neodymium isotopes in the colorless glass of Iulia Felix and Ouest Embiez shipwrecks. Drawing according to [41].

Other Interpretative Diagrams

When more than two parameters influence a particular trait, bi-plot diagrams are not useful. For three independent parameters in observation, triangular (ternary) diagrams are very informative. Fig. (**8**) shows a triangular diagram with apexes TiO_2/Al_2O_3 and Al_2O_3/SiO_2, differentiating high-titanium and low alumina Egyptian sand from high alumina low titanium Levantine sand, criteria introduced by Schibille and colleagues [44], and SrO/CaO criterion used to evaluate amount of strontium that derives from lime. The additional criterion enables differentiation not only between Egyptian and between Levantine sands, but also differentiation of several production groups within Egypt itself: group to which types Foy 3.2 and Foy 2.1 belong, the group to which belong Foy 1, and the Egypt I production group. The type Egypt II is of unknown origin.

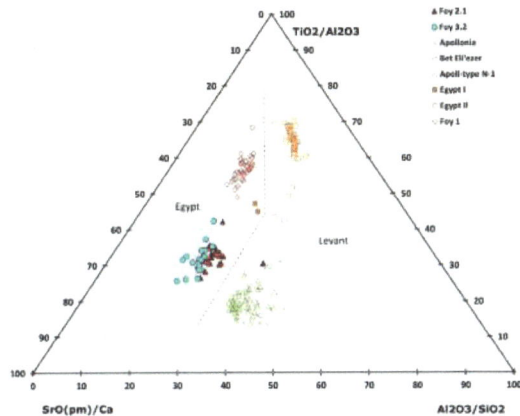

Fig. (8). Ternary diagram of TiO_2/Al_2O_3, $Al_2O_3/SiO2$ and SrO/CaO for Jelica and selected Late Antiquity glass groups, indicative of sand provenance. Ratio MgO/Fe_2O_3 is scaled for better resolution. Data sources: Foy 3.2 and 2.1 [19], Apollonia, Bet Eli'ezer [16], Apollonia N-1, Egypt I, II [18]. SrO given in pm, other values in percent.

When more than three independent variables are determinants of a characteristic in observation, multivariate analysis is the method of choice, the most useful method being the Principal Component Analysis (PCA). The PCA shows general picture of compositional groupings of glasses and their major compositional differences in terms of vectors of oxides concentrations. Fig. (**9**) shows PCA analysis of glass assemblage from Jelica [45] and two late antiquity assemblages, Foy série 3.2 and série 2.1 [19]. Oxides vectors showing heavy sand minerals (MgO, TiO_2) and light sand minerals and lime (Al_2O_3, K_2O, CaO) are oriented roughly to the same direction, indicating the direction of increase of sand impurities. Consequently, the glasses produced with sand containing more impurities plot to the right, those made of cleaner sands to the left. The group on the left, made of cleaner sand, is characterized as Foy 3.2. The glasses with higher concentrations of sand-derived minerals (center, up) is classified as Foy 2.1. These glasses are further divided into two subgroups, spread alongside the iron oxide vector, high iron Foy 2.1 (HI) and very high iron Foy 2.1 (VHI). For distinction, the rest of Foy 2.1 glass is termed low iron Foy 2.1 (LI). The deliberately colored glasses (24 and 33) are depicted in dark blue, its base glass belonging to the Roman glass. The black glass (32) is colored with 6.68% of Fe_2O_3, so it is positioned farther apart from the rest of Foy 2.1 in the direction of the iron oxide vector.

Fig. (9). PCA analysis of Jelica glassware and windowpane glasses, and Foy série 3.2 and Foy série 2.1. Used vectors: Na_2O, MgO, Al_2O_3, SiO_2, K_2O, CaO, TiO_2, MnO, Fe_2O_3. Glasses belonging to Foy 3.2, Foy 2.1 low iron, Foy 2.1 high iron and Foy 2.1 very high iron are outlined with dashed lines. High MgO sub-group is outlined in dotted line. In all 153 glasses are represented with 66.7% of variance explained. Source [29].

However, since PCA takes into accounts many variables, an influence of a particular component that is unrelated to the observed characteristic might move glasses placed on PCA diagram in its direction and give false impression of its adherence. For example, if manganese is not naturally occurring in the glass but is rather deliberately added for decoloring of coloring purposes, then it translates the entire group of glass in the direction of increase of the manganese vector. Other examples include increase of iron in glass from iron blowpipe during glass-blowing [46] or deliberate coloring by iron, as the glass 32 above, increase of alumina from ceramic pots during recycling, or from wall of furnace tanks. In such a case, the affected elements variables might be excluded from PCA diagram, but the problem is that in such a case a variable that is derived from sand is excluded from analysis. One might artificially set value of the particular oxide to average of corresponding composition, but often we do not know to which composition the glass belongs. Additional methods of analysis should be tried out in these cases.

PCA is normally used to graphically display individual components, so it is not suitable to represent large amount of data. In our field of study, this might be the case when there is a need to analyze and compare many glass groups, which have several hundreds of glasses each. The approach to this situation can be based on averages and standard deviations of each group under observation, in the multivariate space of principal components.

The fact that the principal component analysis is a linear method implies that averages and standard deviations of principal components are the same as principal components of averages and standard deviations of individual variables for the dataset in observation. Thus, by calculating averages and standard deviations of the variables in observation first, and performing PCA transformation on them, we can obtain average values and standard deviations in the primary component space and compare multivariate datasets of any size with each other. The example is given in (Fig. **10**), where a large amount of data belonging to many Late Antiquity glass groups from across the Roman Empire is depicted in the single PCA space, enabling their general comparison. In the example, we consider that a sample belongs to a group if it falls within the area defined by the average value plus or minus two standard deviations. This kind of PCA represents a "panorama" of the large datasets enabling us to compare general characteristics, similarities and differences between groups.

The diagram shows that the glass groups are themselves grouped within clusters. If we compare large group of glasses to galaxies, then cluster of groups would correspond to clusters of galaxies. This yields a map of "cosmos" of Late Antiquity glass compositions.

The diagram also shows that using two standard deviations criteria for group adherence leads to a non-negligible overlap between groups. This is because many variables are in the play affecting a position of individual points. This is why PCA is not in every case the exclusive criterion for adherence of a measurement to a particular group. In such cases, other specific criteria need to be implied. For example, to decide adherence to a group in the case when a particular glass falls into the overlapping space belonging to both Roman Mn or Foy 3.2 group (both manganese-decolorized) at the same time, one would need to compare its distinctive features, like the concentrations of sodium or Al_2O_3/SiO_2 and TiO_2/Al_2O_3 sand provenance criteria. However, despite the mentioned limitations, PCA remains the strongest tool for compositional comparison of different glass groups – albeit unpractical for a large amount of data.

For this task, good approach would be to depict only the average value of every group in observation in the PCA space – one point for each group (Fig. **11**). The differences between the groups expressed as positions of average values in the PCA space now become very clear, with having in mind that each group is "condensed" and represented in a single PCA point. The relative "distances" between the groups can now be measured by distances between the points in the PCA space.

The diagram shows three agglomerations. On the left, agglomeration characterized by very high concentrations of elements derived from heavy minerals of the sand, Fe_2O_3, TiO_2, MgO and MnO. The agglomeration consists of the group 1 and series 2.1 of Foy. Compositions similar to the group 1 are termed HIMT1, HIMTa and HIMTb, strong HIMT by different authors. There is a tendency to use the an umbrella-name HIMT for this group. The diagram demonstrates that while Foy 2.1 and Foy 1 are similar, there is a difference between them. Foy 1 is further away from the (0, 0) point in the directions of oxide vectors of heavy minerals. Indeed, Foy 1 has higher concentrations of TiO_2, ZrO_2 and in some cases Fe_2O_3 than Foy 2.1. In addition, these groups characterize higher concentrations of sodium, which is expected for groups produced from Egyptian sands, as explained earlier.

The next agglomeration, the one in the middle-upper space of the diagram, consists of two similar compositions, Roman Sb glass and Foy 3.2. As diagram shows, these types characterizes use of much cleaner sand in raw glass production, with low alumina and high natron. The difference between these two groups consists for the most part, in that Foy 3.2 has manganese instead antimony as decolorized and higher lime concentrations than Roman Sb glass. The diagrams also shows that the early Foy 3.2 glasses ("non-tardifs") were produced with even cleaner sands but with higher lime concentrations compared to "tardifs" Foy 3.2 glasses. In addition, the diagram clearly shows that Foy 3.2 and Foy 2.1 glasses should not be united under the same umbrella name Foy 2, as proposed by some authors in the context when the total number of published glasses of Foy 3.2 type was not large enough to be characterized as a group in its own right [44]. The glass belonging to this agglomeration is also produced in Egypt.

Another agglomeration is positioned bellow the second one, in the direction of increase of lime and decrease of natron. It consists of the Foy group 3.1 and the Levantine I, as termed by Freestone [16]. The relation between 3.1 "non tardifs" and 3.1 is analogous to the relation between the groups 3.2 "non tardifs" and 3.2. The glass from this agglomeration is produced in the Levant.

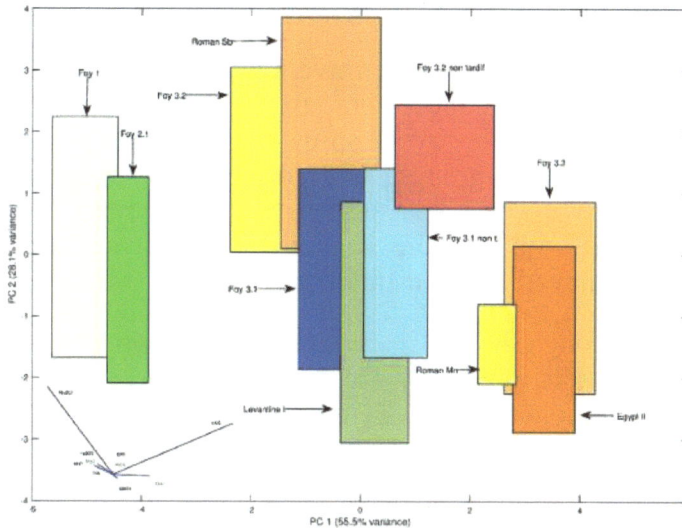

Fig. (10). PCA diagram depicting averages plus/minus standard deviations of various groups of natron glass of the Late Antiquity in the multivariate space, as rectangles. Oxide vectors are superimposed, with coordinate center in the point (0,0) of the PCA diagram. Glass groups shown: Roman Sb, Roman Mn and Levantine I of Schibille and colleagues [44], Foy 1, 2.1, 3.1, 3.2 and 3.3 of Foy and colleagues [19], Levantine I of Freestone [16], Egypt II of Phelps and colleagues [18].

The fourth agglomeration occupies the far right of the PCA diagram, in the direction of lime and alumina vectors, and to the opposite direction of vectors showing increase of elements derived from heavy minerals, like iron, titanium and magnesium. The position on the diagram shows use of sand low in heavy minerals and high in feldspars and lime. It consists of the groups Roman Mn, Foy 3.3 (Levantine II in Freestone's terminology) and the group Egypt II [18]. Apart from the Egypt II, whose provenance is unknown, the glasses belonging to these groups is considered of Levantine origin. Thus, the PCA diagram represents a map in the multivariate space of glass-making sands with differing regional geology, together with some technological characteristics, such as natron/sand recipes and the deliberate use of manganese to decolorize the glass.

To analyze only sand characteristics, flux and coloring-related elements can be excluded from the diagram, but in such a case one should take into account the fact that manganese is also a natural component of sand, especially so for the sands with higher concentrations of heavy minerals, like HIMT glasses.

PCA diagrams can be used in yet another way, if there is a need to numerically assess the overall compositional difference between two large glass groups, especially in the case when they are overlapping. To do this, the distance between the "centers of compositional gravity" of each group, represented by the average composition in the multivariate PCA space, can be taken as a measure of such difference. Fig. (11) shows the PCA diagram of the same groups as in (Fig. 10), here represented by single points – their average composition in the space of two principal components. The distances demonstrate that the Roman Sb and the Foy 3.2, albeit with considerable overlapping in the regular PCA space, are nevertheless different glass groups, with distance between their "centers of gravity" equal to 1.77 in the space of first two primary components. The conclusion is that Foy 3.2 and the Roman Sb (albeit similar, both being of Egyptian origin) are nevertheless different.

On the other side, although there is no too much overlapping between Foy 3.1 and Levantine I, their centers of gravity are only 0.89 apart from each other in the space of principal components 1 and 2, indicating that they nevertheless represent a single compositional group. This approach in some cases might provide an instrument to judge about conflicting naming conventions, as is the case here where different authors termed their own respective glass assemblages.

Time Diagrams

Until now, we have analyzed the assemblages from a specified, fixed period. However, if certain production group lasted for a longer period (and this is the case of many groups, like Roman Sb, Foy 3.2, HIMT), diachronic analysis is in principle possible. Fig. (12) (up) depicts how alumina concentration was changing in Foy 3.2 type of glass, in the long period from the second to the sixth century CE. In the period from the $2^{nd} - 5^{th}$ century, alumina concentrations have a decreasing trend, showing that sands with less feldspars were used. In the 6th century, trend reverses. The Fig. (12) (middle, lower) shows that manganese concentrations were in constant rise, apart from the 4th to the 5th century, which might be an indication that in this period cleaner sand was used or market taste changed for the less decolorized glass.

Fig. (12) (down) shows that in the 3rd century a fair amount of Foy 3.2 glass was recycled with antimony-containing cullet, and this drops almost to zero in the 4th century, when manganese replaced antimony as the main decolorizer making antimony-containing cullet less available. Thus, Fig. (12) reflects changes in composition that derive from changing glass-making sands, but can also reflect some other factors, like market forces.

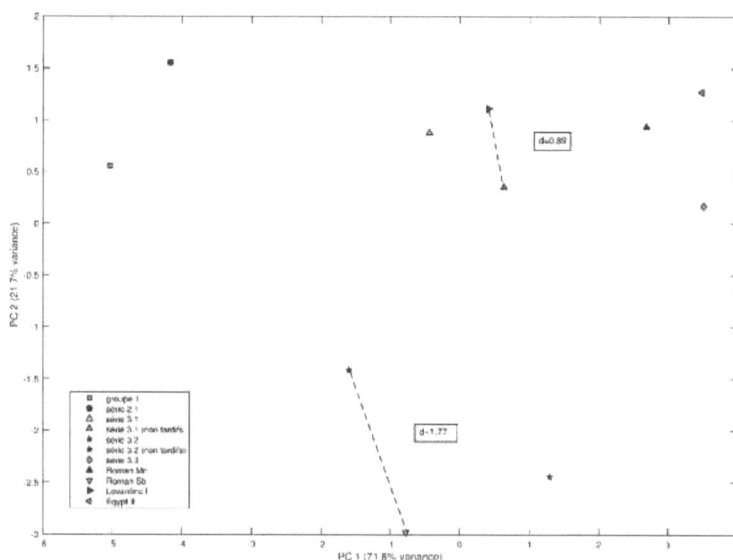

Fig. (11). Principal component analysis of the average compositional values of particular groups. Distances between centers of Foy 3.2 and Roman Sb group (lower in the diagram) and between Foy 3.1 and Levantine I (upper) are shown. Groups represented are the same as in the (Fig. **10**).

Histograms and Bar Graphs

However simple, histograms and bar graphs can reveal important features of glass distributions and glass usage and provide bases for interpretations. If a particular type of glass is used in particular types of objects and not in other types, or in particular place or a defined time-span, it can reveal patterns of its use, distribution or time evolution of glass markets. The example Fig. (**13**) (up) shows the distribution of Foy 2.1 with elevated concentrations of iron (Fe-rich glasses) in different parts of the wider Mediterranean region. The higher concentration in the regions closer to Egypt, like Cyprus, might indicate that this type was produced mainly in Egypt. The distribution of Fe-rich glass among the types of glass objects (Fig. **13** below) might be interpreted so that not all iron was added to glass in primary workshops but also in the secondary workshops during glass blowing or glass casting. This might further imply a different origin of iron in these two cases. In addition, since higher iron concentrations mean darker green glass, the higher percentage of Fe-rich glass among glassware than in windowpane may be interpreted as a decorative feature of glassware.

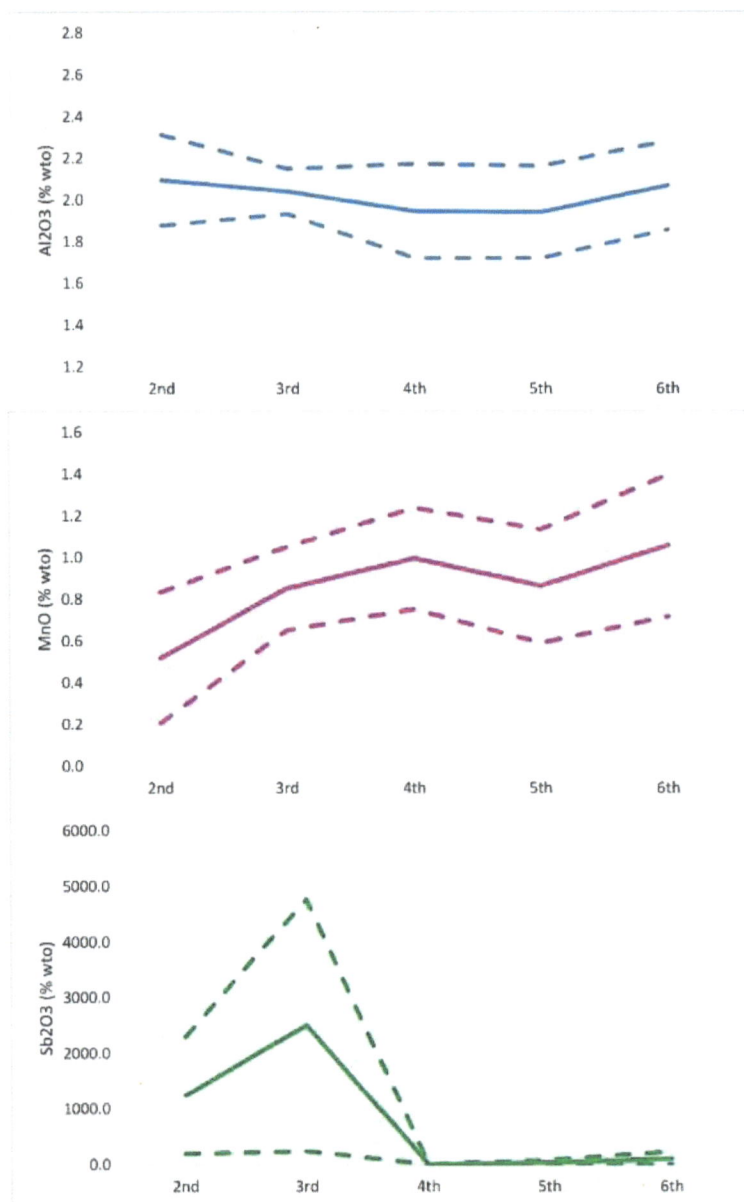

Fig. (12). Alumina (up), manganese (middle), and antimony (lower) average and standard deviations of 195 glasses of Foy 3.2 type from the literature, belonging to 21 different glass groups across the Mediterranean and Western Europe, 2nd -6th c. CE. Averages – straight lines, standard deviations – dotted lines.

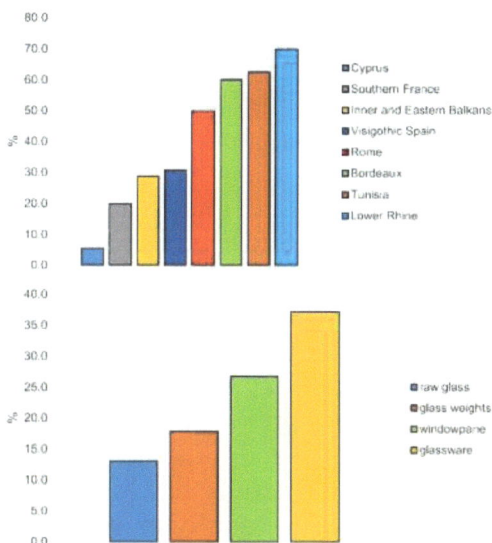

Fig. (13). Percentages of Fe-rich glasses by region and type. Percentages of Fe-rich glasses among Foy 2.1 glasses by the region (**a**); Percentages of Fe-rich glasses among Foy 2.1 glasses by object type (**b**). Data sources: [19, 29, 45 - 47, 34, 48 - 51].

Statistical Data Analysis

Another useful method is the Kernel density estimate (KDE) [52, 53]. In order to differentiate eventual subgroups of the Foy 2.1 group with elevated iron (Fe-rich Foy 2.1), 134 glasses from eleven sites published in the literature are collected and fitted (Fig. **14**). The ratio of Fe_2O_3/TiO_2 is taken as an indicator of heavy minerals in sand. It seems that there exist at least two groups regarding iron-to-titanium ratios.

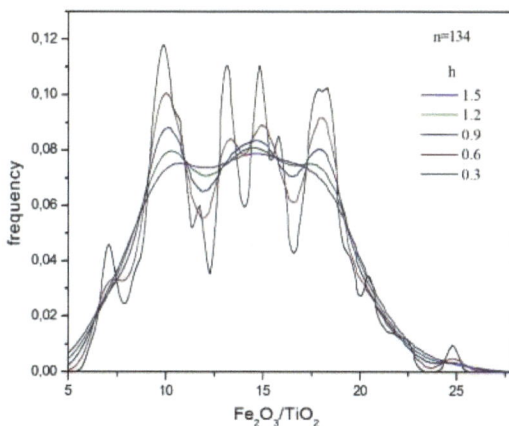

Fig. (14). Kernel density estimate for the ratio Fe_2O_3/TiO_2 for 134 glasses of Fe-rich Foy 2.1 type. The calculation was performed for five different bandwidth parameters h.

The fitted curves intersect at the Fe_2O_3/TiO_2 value of approximately 11.8. The diagrams show a distinct possibility that a third group also exists, but more data is needed to make definitive conclusions.

Artificial Intelligence in Data Analysis

In principle, some of the questions of the interpretative field might be approached by artificial intelligence methods. Since a typical task for expert systems is classification, perhaps the question of adherence of a particular glass sample to a glass type or a group might be solved (or at least eased and speeded up) by an expert system. The knowledge base can be formed from the already established expert knowledge. For example, a rule in the knowledge base can be the mentioned criteria introduced by Schibille and colleagues that discriminate the Egyptian sand from the Levantine sand. Another example can be the heuristic knowledge that glass from Egypt generally has higher concentrations of sodium due to the proximity of Wadi Natrun, usually more than 18 percent. Yet another rule can be established that Sr/CaO ratio in the range from 40 to 60 is the approximate value for the Eastern Mediterranean beach sands. The inference engine of the expert system would then interpret these rules in order to answer the question of the glass provenance, whether it is Egypt or Levantine cost. Production rules can be defined to form a more extensive knowledge base. Such system, applied on the database of glass compositions, can speed up the compositional characterizations of large data sets.

Other artificial intelligence areas might be used for other purposes. Perhaps pattern recognition techniques might be used in the recognition of fragments of glass, which are usually found in archaeological excavations. Such an application, used perhaps as a screening method, would greatly speed up the procedure of determining the type of glass object where a bigger number of fragments is encountered. Perhaps the AI technique might even be able to offer several possible reconstructed shapes to aid an expert or to reconstruct a shape that is too small for reconstruction by eye. It is understood that an expert would always take the final decision.

CONCEPTUALIZATION AND INTERPRETATION OF DATA

We have already touched several concepts covering the space between archaeology and physics. We will now try to elaborate and systematize them further. We start by naming the area between the questions posed by archeology and the analytical methods of physics, the interpretative field. The interpretative field is a heterogenous set of approaches and concepts based on heuristics. In our case study of archaeometry of glass, the interpretative field comprises three groups of concepts: the group of technological concepts, the group of classifying concepts, and the group of archaeological concepts.

Technological Concepts

Concepts in this group relate to the manufacturing practices in ancient glass production. One of them is the type of flux used in melting sand in order to reduce the melting temperature. During the observed period, flux was mostly mineral natron but also halophytic plant-ash. Later on, flux from trees and bracken, called wood-ash flux, appeared. Thus, we have several types of ancient glass regarding the flux used: natron glass, plant-ash glass and wood-ash glass.

Another technological concept relates to the stabilization of the glass. This was done in several ways, choosing the sand naturally rich in lime, using halophytic plants for fluxing, or adding sea shells.

Yet another important concept, derived from archaeological and scientific observations, is the concept of the two-phased organization of glass production. First the "raw glass" was produced near sources of good quality and abundant reserves of raw materials and energy used in glass production (sand, natron, wood for firing and plants for ashing). In the second phase, raw glass was melted again in other workshops, and often mixed with glass cullet.

The concept of recycling is another major technological concept. Melting in the secondary workshops often included cullet. The second melting was an appropriate time to give a future glass object qualities like color and opacity. Concepts that deal with the relevant technological processes are coloring, decoloring and opacifying [54]. It seems that coloring was not used solely for decorative purposes, but also for marketing purposes. The color-branding hypothesis [55] states that the HIMT glass was intentionally colored with manganese to brand with color its excellent glass-working characteristics. Different manganese ores were supposedly used according to the hypothesis, yielding different types of HIMT glass, HIMTa and HIMTb.

Classifying Concepts

Classifying concepts group glasses in observation according to some criteria or attribute. Since these criteria or attributes are more than a few, the classifying concepts are sometimes ambiguous, overlapping and even contradictory. Naming conventions are often coupled to these classifications, generating sometimes complicated and fuzzy nomenclature. Authors mostly start by naming the collection of glass from a particular archaeological site S, and then grouping glasses with similar composition numerically: S1, S2, *etc.* When new assemblages of glass are reported, each with its own name and grouping, authors try to find similarities between the groups. The problem arises when more and more assemblages are reported, and soon we end up with intertwined and confusing set of names. These groups are compositionally rarely the same, and they quite often overlap.

The concept of "type" of glass takes compositional specifics as criteria to define a particular type. The names given to types are mostly explanatory (Roman colorless Sb, Roman colorless Mn, Roman naturally colored, *etc.*) or their acronyms (HIMT – high iron, manganese, titanium). Foy and colleagues systematized their big collection of glass into compositional groups and subgroups (*series* in their terminology). A significant effort in naming systematization is given by Gliozzo, with her coding system of classification of the Roman and the Late Roman colorless and colored glass.

However, all the mentioned criteria are "static" in a sense that they do not take into account the passage of time during the lifetime of any particular glass type. As we have seen earlier in this text, glassmakers were in a constant quest for ever-new sources of raw material and fuel. As the archaeological data show, the primary glass workshop moved to another place after a few years of exploitation, typically two or three. This means the constant change in sand composition reflected itself in small changes of glass composition within a single glass type (Fig. 12). Therefore, fluctuations in the composition of glass of any glass type are expected. In addition, the unceasing quest for new raw materials and incremental movements of a glass quarry might have led to a geological area with notably different geochemistry of sand. For this reason, the criteria to be used to determine if one compositional type changed to another need to be connected to the understanding of the geochemistry of glass-making sands in question. We call this type of classification, based on the geology of sand, the "geological class".

Archaeological Concepts

Archaeological concepts as treated here are particularization of general questions posed by archeology: what, when and how something appeared, which forms did it take and what interaction with society did it have? These general questions are then decomposed to specific questions such as what is the dating, what is the type of an object, what was its use, who used it and how, how was it produced, how was it transported, what was its impact? The archaeological concepts are built upon these questions. In the field of ancient glass, they translate to the concepts of typology of glass objects and glass compositions, dating, technology of production, organization of production, organization of trade, glass-trading networks and trading patterns, marketing, market forces, and impacts on society.

This step towards formalization of the interpretative field is shown in (Fig. 15). It is important to note that the flow of information between three groups of concepts is a two-way street. Table **1** summarizes the interpretative field concepts.

Table 1. The interpretative field consists of four groups of concepts, emerging from two areas at the same time of archeology and the field of physical measurements. The table is given in relation to archaeological glass.

Analytical Dependences	Technological Concepts	Classifying Concepts	Archaeological Concepts
Bi-plots – two parameters	Fluxing agent – natron, plant ash, wood ash	Assemblage – heterogeneous selection of glasses	Dating
Triangular diagrams – three parameters	Two-phased production – primary and secondary workshops	Type – composition with particular concentrations of oxides within assemblage	Typology
PCA diagrams – many parameters	Stabilization – adding lime or selecting sand with high CaO	Compositional group - compositions with a particular concentration of oxides generally	Production organization
Correlations – indication of minerals, recycling	Recycling – using broken glass, cullet	Series – compositional subdivision within the group	Trade networks
Time diagrams – change of composition in time	Coloring – furnace conditions, ion oxidation states	Production group - composition with particular concentrations of oxides related to the primary workshops	Trading patterns
Histograms – distributions of variables	Decoloring – antimony, manganese	Geological class - composition with particular concentrations of oxides related to the geological area	Marketing
Bar graphs – distributions	Opacifying – antimonates	-	Impacts on society
REE patterns – provenance	Color-branding – manganese ore	-	-
Isotopic ratios – provenance	Batch [56] - glass produced in one firing of furnace; proportional compositional differences	-	-

Fig. (**15**) also depicts the flow of information between the analysts and the archaeologists. The flow goes through the area we call the Interpretative Field, using sets of analytical diagrams, and three sets of concepts - technological, classifying and archaeological. This flow is iterative, with each cycle of discussions gaining more precise answers to the posed questions. The yellow star represents this iterative communication, where each arrow represents a person participating in the analysis, natural scientist or archaeologist. While it is not necessary (nor possible) that a scientist nor an archaeologist know the science of each other, it is helpful that they understand and use the concepts that lie in between, which we term the Interpretative Field and which comprises several sets of heuristic but well-defined and clearly understood concepts.

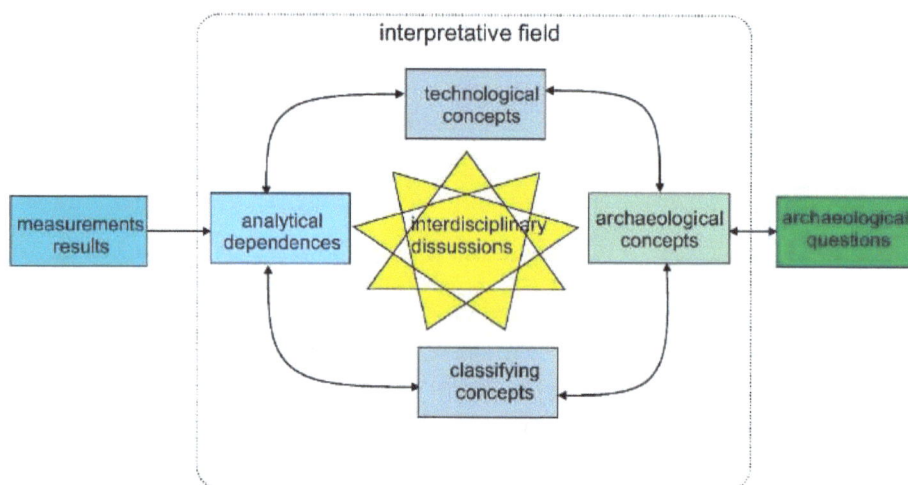

Fig. (15). Schematic of the flow of information between the physicist and the archaeologist during their interdisciplinary work. Iterative discussions are through the sets of heuristic concepts that we term the Interpretative Field.

CONCLUSION

Central to the meaningful and reliable interpretation of the results of scientific measurements undertaken to answer questions posed by social sciences, like archeology, are clearly and consistently defined concepts and notions deriving from both ends of this scientific communication – natural and social sciences. How are these concepts defined, what is their level of abstraction, what is their syntax and semantics, and how they relate to each other; all these questions fall into an area that we dub "interpretative field". The interpretative field is the place that generates the answers to questions posed by social sciences, like archeology, to exact sciences, like physics or chemistry. We propose that this field be as well-defined and as abstract as possible with the idea that it would be easier to understand and grasp by both the natural and the social scientist. We illustrated this approach in the case study of PIXE/PIGE measurements of archaeological glass, demonstrating many ways the measurements are treated and interpreted in the context of the questions asked by archaeology.

REFERENCES

[1]   Pichon, L.; Moignard, B.; Lemasson, Q.; Pacheco, C.; Walter, P. Development of a multi-detector and a systematic imaging system on the AGLAE external beam. *Nucl. Instrum. Methods Phys. Res. B,* **2014**, *318*, 27-31.
[http://dx.doi.org/10.1016/j.nimb.2013.06.065]

[2]   Topić, N.; Radović, I.B.; Fazinić, S.; Šmit, Ž.; Sijarić, M.; Gudelj, L.; Burić, T. Compositional analysis of Late Medieval glass from the western Balkan and eastern Adriatic hinterland. *Archaeol. Anthropol. Sci.,* **2019**, *11*(5), 2347-2365.
[http://dx.doi.org/10.1007/s12520-018-0712-9]

[3]   Šmit, Ž. Physics beyond PIXE. *Nucl. Instrum. Methods Phys. Res. B,* **2020**, *477*, 13-18.
[http://dx.doi.org/10.1016/j.nimb.2019.09.019]

[4]   Šmit, Ž.; Uršič, M.; Pelicon, P.; Trček-Pečak, T.; Šeme, B.; Smrekar, A.; Langus, I.; Nemec, I.; Kavkler, K. Concentration profiles in paint layers studied by differential PIXE. *Nucl. Instrum. Methods Phys. Res. B,* **2008**, *266*(9), 2047-2059.
[http://dx.doi.org/10.1016/j.nimb.2008.03.191]

[5]   Chen, C.; Freestone, I.C.; Gorin-Rosen, Y.; Quinn, P.S. A glass workshop in 'Aqir, Israel and a new type of compositional contamination. *J. Archaeol. Sci. Rep.,* **2021**, *35*, 102786.
[http://dx.doi.org/10.1016/j.jasrep.2020.102786]

[6]   Sayre, E.V.; Smith, R.W. Compositional categories of ancient glass. *Science,* **1961**, *133*(3467), 1824-1826.
[http://dx.doi.org/10.1126/science.133.3467.1824] [PMID: 17818999]

[7]   Rosenow, D.; Rehren, Th. A view from the south: Roman and late antique glass from Armant, Upper Egypt. In Rosenow, D.; Phelps, M.; Meek, A.; Freestone, I.C. (Eds,) Things that Travelled: Mediterranean Glass in the First Millennium AD. UCL Press, London **2018**, 283-323.

[8]   Van Ham-Meert, A.; Claeys, P.; Jasim, S.; Overlaet, B.; Yousif, E.; Degryse, P. Plant ash glass from first century CE Dibba, U.A.E. *Archaeol. Anthropol. Sci.,* **2019**, *11*(4), 1431-1441.
[http://dx.doi.org/10.1007/s12520-018-0611-0]

[9]   Brill, R.H. Scientific investigations of Jalame glass and related finds, Ed. Weinberg G.D., Excavations in Jalame: Site of a Glass Factory in Late Roman Palestine. University of Missouri, Columbia **1988**, 257-294.

[10]   Henderson, J. Tradition and experiment in first millennium a.d. glass production--the emergence of early Islamic glass technology in late antiquity. *Acc. Chem. Res.,* **2002**, *35*(8), 594-602.
[http://dx.doi.org/10.1021/ar0002020] [PMID: 12186563]

[11]   Balvanović, R.; Šmit, Ž.; Stojanović, M.M.; Spasić-Đurić, D.; Špehar, P.; Milović, O. Late Roman glass from Viminacium and Egeta (Serbia): glass-trading patterns on Iron Gates Danubian Limes. *Archaeol. Anthropol. Sci.,* **2022**, *14*(4), 79.
[http://dx.doi.org/10.1007/s12520-022-01529-y]

[12]   Mirti, P.; Casoli, A.; Appolonia, L. Scientific analysis of Roman glass from Augusta Praetoria. *Archaeometry,* **1993**, *35*(2), 225-240.
[http://dx.doi.org/10.1111/j.1475-4754.1993.tb01037.x]

[13]   Freestone, I.C. Appendix: Chemical Analysis of 'raw' glass fragments. In: Hurst, H.R. (Ed.), Excavations at Carthage, Vol. II, 1. The Circular Harbour, North Side. The Site and Finds other that Pottery. British Academy Monographs in Archeology, No. 4. Oxford University Press, Oxford **1994**, *290*

[14]   Foster, H.E.; Jackson, C.M. The composition of 'naturally coloured' late Roman vessel glass from Britain and the implications for models of glass production and supply. *J. Archaeol. Sci.,* **2009**, *36*(2), 189-204.
[http://dx.doi.org/10.1016/j.jas.2008.08.008]

[15] Freestone, I.C.; Wolf, S.; Thirlwall, M. The production of HIMT glass: elemental and isotopic evidence. Annales du 16e Congre' s de l'Association Internationale pour l'Histoire du Verre. London, UK. **2005**, , 153-157.

[16] Freestone, I.C.; Gorin-Rosen, M.Y.; Hughes, M.J. Freestone, I. C.; Gorin-Rosen, M. Y.; Hughes, M. J. Primary glass from Israel and the Production of Glass in Late Antiquity and the Early Islamic Period, La Route du verre, Lyon, **2000**, 65-83.

[17] Gratuze, B.; Barrandon, J.N. Islamic glass weights and stamps: Analysis using nuclear techniques. *Archaeometry,* **1990**, *32*(2), 155-162.
[http://dx.doi.org/10.1111/j.1475-4754.1990.tb00462.x]

[18] Phelps, M.; Freestone, I.C.; Gorin-Rosen, Y.; Gratuze, B. Natron glass production and supply in the late antique and early medieval Near East: The effect of the Byzantine-Islamic transition. *J. Archaeol. Sci.,* **2016**, *75*, 57-71.
[http://dx.doi.org/10.1016/j.jas.2016.08.006]

[19] Foy, D.; Picon, M.; Vichy, M.; Thirion-Merle, V. Caractérisation des verres de la fin de l'Antiquité en Méditerranée occidentale : l'émergence de nouveaux courants commerciaux, in: "Échanges et commerce du verre dans le monde Antique", Aix-en-Provence et Marseiile, Montagnac, Ed. M. Mergoil (Monographies Instrumentum 24) **2003**, 41-86.

[20] Foy, D.; Vichy, M.; Picon, M. Lingots de verre en Mediterranee occidentale (IIIe siècle av. J.-C. – VIIe siècle ap. J.-C.), in Annales du 14e Congrès de l'Association Internationale pour l'Historie du Verre, 2000, Italia/Venezia–Milano, 27 October – 1 November 1998, 51–7, AIHV, Lochem **2000**.

[21] Freestone, I.C. Primary glass sources in the mid first millennium AD in Annales du 15e Congrès de l'Association Internationale pour l'Histoire du Verre New York – Corning 15 – 20 111–5 AIHV, 2003, Nottingham.

[22] Freestone, I.C. The Provenance of Ancient Glass through Compositional Analysis **2004**.
[http://dx.doi.org/10.1557/PROC-852-OO8.1]

[23] Nenna, M.D.; Picon, M.; Vichy, M. Ateliers primaires et secondaires en Egypte à l'époque grécoromaine La route du verre TMO 33. **2000**.

[24] Nenna, M. D.; Picon, M.; Thirion-Merle, V.; Vichy, M. Nenna, M. D. ; Picon, M. ; Thirion-Merle ,V. ; Vichy, M. Ateliers primaries du Wadi Natrun: nouvelles découvertes AIHV Annales du 16e Congrés **2003**, 59-63.

[25] Gorin-Rosen, Y. The Ancient Glass Industry in Israel: Summary of the Finds and New Discoveries La Route du verre. Ateliers primaires et secondaires du second millénaire av. J.-C. au Moyen Âge. Colloque organisé en 1989 par l'Association française pour l'Archéologie du Verre (AFAV) Lyon: Maison de l'Orient et de la Méditerranée Jean Pouilloux **2000**, 49-63.

[26] Balvanović, R; Šmit, Ž Emerging Glass Industry Patterns in Late Antiquity Balkans and Beyond: New analytical Findings on Foy 3.2 and Foy 2.1 Glass Types, Materials **2022**.

[27] Paynter, S.; Jackson, C. Clarity and brilliance: antimony in colourless natron glass explored using Roman glass found in Britain. *Archaeol. Anthropol. Sci.,* **2019**, *11*(4), 1533-1551.
[http://dx.doi.org/10.1007/s12520-017-0591-5]

[28] Stamenković, S. Greiff, S,; Hartmann, S.; Late Roman glass from Mala Kopašnica (Serbia) - forms and chemical analysis, AIHV Annales du 20e Congrès **2015**.

[29] Balvanović, R.; Šmit, Ž. Sixth-century AD glassware from Jelica, Serbia—an increasingly complex picture of late antiquity glass composition. *Archaeol. Anthropol. Sci.,* **2020**, *12*(4), 94.
[http://dx.doi.org/10.1007/s12520-020-01031-3]

[30] Jackson, C. M. Making colourless glass in the Roman period Archaeometry 47, 4 **2005**, 763-780.

[31] Culicov, O.A.; Trtić-Petrović, T.; Balvanović, R.; Ražić, S.S.; Petković, A. Spatial Distribution of multielements including lanthanides in sediments of Iron Gate I Reservoir on the Danube River.

Environ. Sci. Pollut. Res. Int., **2020**.
[PMID: 33851297]

[32] Schibille, N.; Meek, A.; Tobias, B.; Entwistle, C.; Avisseau-Broustet, M.; Da Mota, H.; Gratuze, B. Comprehensive chemical characterisation of Byzantine glass weights, Plos One, 2016. **2016**.
[http://dx.doi.org/10 1371/journal pone 0168289]

[33] Ceglia, A.; Cosyns, P.; Nys, K.; Terryn, H.; Thienpont, H.; Meulebroeck, W. Ceglia, A.; Cosyns, P.; Nys, K.; Terryn, H.; Thienpont, H. Meulebroeck, W. Late antique glass distribution and consumption in Cyprus: a chemical Study Journal of Archaeological Science 61 213-222 2015; doi: 10 1016/j jas, 2015, 06 009 **2015**.
[http://dx.doi.org/10 1371/journal pone 0168289]

[34] Ceglia, A.; Cosyns, P.; Schibille, N.; Meulebroeck, W. Unravelling provenance and recycling of late antique glass from Cyprus with trace elements. *Archaeol. Anthropol. Sci.,* **2019**, *11*(1), 279-291.
[http://dx.doi.org/10.1007/s12520-017-0542-1]

[35] Kamber, B.S.; Greig, A.; Collerson, K.D. A new estimate for the composition of weathered young upper continental crust from alluvial sediments, Queensland, Australia. *Geochim. Cosmochim. Acta,* **2005**, *69*(4), 1041-1058.
[http://dx.doi.org/10.1016/j.gca.2004.08.020]

[36] Ares, J.J.; Guirado, V. E.; Gutiérrez, Y. C.; Schibille, N. Ares, J.J.; Guirado, V. E.; Gutiérrez, Y. C.; Schibille, N. Changes in the supply of eastern Mediterranean glasses to Visigothic Spain, Journal of Archaeological Science 107, 2019 **2019**, 23-31.

[37] Weldeab, S.; Emeis, K.C.; Hemleben, C.; Siebel, W. Provenance of lithogenic surface sediments and pathways of riverine suspended matter in the Eastern Mediterranean Sea: evidence from 143Nd/144Nd and 87Sr/86Sr ratios. *Chem. Geol.,* **2002**, *186*(1-2), 139-149.
[http://dx.doi.org/10.1016/S0009-2541(01)00415-6]

[38] Jackson, C.M.; Paynter, S.; Nenna, M.D.; Degryse, P. Glassmaking using natron from el-Barnugi (Egypt); Pliny and the Roman glass industry. *Archaeol. Anthropol. Sci.,* **2018**, *10*(5), 1179-1191.
[http://dx.doi.org/10.1007/s12520-016-0447-4]

[39] Brems, D.; Ganio, M.; Latruwe, K.; Balcaen, L.; Carremans, M.; Gimeno, D.; Silvestri, A.; Vanhaecke, F.; Muchez, P.; Degryse, P. Isotopes on the beach, Part 1: Strontium isotope ratios as a provenance indicator for lime raw materials used in Roman glassmaking. *Archaeometry,* **2013**, *55*(2), 214-234.
[http://dx.doi.org/10.1111/j.1475-4754.2012.00702.x]

[40] Brems, D.; Ganio, M.; Latruwe, K.; Balcaen, L.; Carremans, M.; Gimeno, D.; Silvestri, A.; Vanhaecke, F.; Muchez, P.; Degryse, P. Isotopes on the beach, Part 2: Neodymium isotopic analysis for the provenancing of Roman glass-making. *Archaeometry,* **2013**, *55*(3), 449-464.
[http://dx.doi.org/10.1111/j.1475-4754.2012.00701.x]

[41] Ganio, M.; Boyen, S.; Brems, D.; Scott, R.; Foy, D.; Latruwe, K.; Molin, G.; Silvestri, A.; Vanhaecke, F.; Degryse, P. Trade routes across the Mediterranean, a Sr/Nd isotopic investigation of Roman colourless glass. *Glass Technol.,* **2012**, *53*, 217-224.

[42] Barfod, G.H.; Freestone, I.C.; Lesher, C.E.; Lichtenberger, A.; Raja, R. 'Alexandrian' glass confirmed by hafnium isotopes. *Sci. Rep.,* **2020**, *10*(1), 11322.
[http://dx.doi.org/10.1038/s41598-020-68089-w] [PMID: 31913322]

[43] Henderson, J.; Ma, H.; Evans, J. Glass production for the Silk Road? Provenance and trade of islamic glasses using isotopic and chemical analyses in a geological context. *J. Archaeol. Sci.,* **2020**, *119*, 105164.
[http://dx.doi.org/10.1016/j.jas.2020.105164]

[44] Schibille, N.; Sterrett-Krause, A.; Freestone, I. C. Schibille, N.; Sterrett-Krause, A.; Freestone, I. C. Glass groups glass supply and recycling in late Roman Carthage Archeological and Antrhropological Sciences **2016**, *9*(6), 1223-1241.

[45]   Balvanović, R.; Stojanović, M.M.; Šmit, Ž. Exploring the unknown Balkans: Early Byzantine glass from Jelica Mt. in Serbia and its contemporary neighbours. *J. Radioanal. Nucl. Chem.,* **2018**.

[46]   Schibille, N.; Freestone, I.C. Composition, production and procurement of glass at San Vincenzo Al Volturno: an early Medieval monastic complex in Southern Italy. *PLoS One,* **2013**, *8*(10), e76479.
[http://dx.doi.org/10.1371/journal.pone.0076479] [PMID: 24146876]

[47]   Drauschke, J.; Greiff, S. Early Byzantine glass from Caričin Grad/Iustiniana Prima (Serbia): first results concerning the composition of raw glass chunks. *Glass along the silk road from 2000 BC to AD 1000.,* **2010**, , 53-67.

[48]   Schibille, N.; Meek, A.; Tobias, B.; Entwistle, C.; Avisseau-Broustet, M.; Da Mota, H.; Gratuze, B. Comprehensive chemical characterisation of Byzantine glass weights. *PLoS One,* **2016**, *11*(12), e0168289.
[http://dx.doi.org/10.1371/journal.pone.0168289] [PMID: 27959963]

[49]   Mirti, P.; Lepora, A.; Saguì, L. Scientific analysis of fragments from the seventh-century glass Crypta Balbi in Rome. *Archaeometry,* **2000**, *42*(2), 359-374.
[http://dx.doi.org/10.1111/j.1475-4754.2000.tb00887.x]

[50]   Velde, B. Aluminum and calcium oxide content of glass found in western and northern Europe, first to ninth centuries. *Oxf. J. Archaeol.,* **1990**, *9*(1), 105-117.
[http://dx.doi.org/10.1111/j.1468-0092.1990.tb00218.x]

[51]   Wedepohl, K.H.; Pirling, R.; Hartmann, G. Römische und fränkische Glaäser aus dem Gräberfeld von Krefeld-Gellep. *Bonner Jahrbücher,* **1997**, *197*, 177-189.

[52]   Juan Ares, J.; Schibille, N.; Vidal, J.M.; Sánchez de Prado, M.D. The supply of glass at Portus Illicitanus (Alicante, Spain): A meta-analysis of HIMT glasses. *Archaeometry,* **2019**, *61*(3), 647-662.
[http://dx.doi.org/10.1111/arcm.12446] [PMID: 31244490]

[53]   Baxter, M.J.; Buck, C.E. Data handling and statistical analysis. *Modern analytical methods in arts and archaeology.,* **2000**, , 681-746.

[54]   Schibille, N.; Neri, E.; Ebanista, C.; Ammar, M.R.; Bisconti, F. Something old, something new: the late antique mosaics from the catacomb of San Gennaro (Naples). *J. Archaeol. Sci. Rep.,* **2018**, *20*, 411-422.
[http://dx.doi.org/10.1016/j.jasrep.2018.05.024]

[55]   Freestone, I.C.; Degryse, P. Lankton. J.; Gratuze, B.; Schneider, J. HIMT, glass composition and the commodity branding in the primary glass industry. *Things that travelled, Mediterranean Glass in the First Millennium CE.,* **2018**, , 159-190.
[http://dx.doi.org/10.2307/j.ctt21c4tb3.14]

[56]   Freestone, I.C.; Price, J.; Cartwright, C.R.C. The batch: its recognition and significance. *Annales du 17e Congrès de l'Association Internationale pour l'Histoire du Verre,* Anvers**2006**, 130-5.

SUBJECT INDEX